THE INFLUENCE OF SEWAGE SLUDGE APPLICATION ON PHYSICAL AND BIOLOGICAL PROPERTIES OF SOILS

Commission of the European Communities

THE INFLUENCE OF SEWAGE SLUDGE APPLICATION ON PHYSICAL AND BIOLOGICAL PROPERTIES OF SOILS

*Proceedings of a Seminar organized jointly by
the Commission of the European Communities, Directorate-General for Science,
Research and Development and the Bayerische Landesanstalt für Bodenkultur
und Pflanzenbau, Munich, Federal Republic of Germany,
held in Munich, June 23–24, 1981*

Edited by

G. CATROUX
Ministère de l'Agriculture, INRA, Laboratoire de Microbiologie des Sols, Dijon, France

P. L'HERMITE
Commission of the European Communities, Brussels, Belgium

and

E. SUESS
Bayerische Landesanstalt für Bodenkultur und Pflanzenbau, Munich, F.R.G.

D. REIDEL PUBLISHING COMPANY
DORDRECHT : HOLLAND / BOSTON : U.S.A.
LONDON : ENGLAND

Library of Congress Cataloging in Publication Data

Main entry under title:

The influence of sewage sludge application on physical and biological properties
    of soils.

At head of title: Commission of the European Communities.
Papers presented at the Seminar on "The Influence of Sewage Sludge
Application on Physical and Biological Properties of Soils", Munich, Germany,
June 23–24, 1981.
    1. Sewage sludge as fertilizer—Congresses. 2. Sewage sludge –
Environmental aspects—Congresses. 3. Soil physics—Congresses. 4. Soil
biology—Congresses. I. Catroux, G., 1938–   . II. L'Hermite, P.
(Pierre), 1936–   . III. Suess, E. IV. Commission of the European
Communities. V. Seminar on "The Influence of Sewage Sludge Application
on Physical and Biological Properties of Soils" (1981 : Munich, Germany)
S657.I54   1982   631.8'69   82-15106
ISBN-13:978-94-009-7932-1   e-ISBN-13:978-94-009-7930-7
DOI: 10.1007/978-94-009-7930-7

Publication arrangements by
Commission of the European Communities
Directorate-General Information Market and Innovation, Luxembourg

EUR 8023
Copyright © 1983 ECSC, EEC, EAEC, Brussels and Luxembourg
Softcover reprint of the hardcover 1st edition 1983

LEGAL NOTICE

Published by D. Reidel Publishing Company
P.O. Box 17, 3300 AA Dordrecht, Holland

Sold and distributed in the U.S.A. and Canada
by Kluwer Boston Inc.,
190 Old Derby Street, Hingham, MA 02043, U.S.A.

In all other countries, sold and distributed
by Kluwer Academic Publishers Group,
P.O. Box 322, 3300 AH Dordrecht, Holland

D. Reidel Publishing Company is a member of the Kluwer Group

C O N T E N T S

Preface

SESSION 1 - EFFECTS OF SEWAGE SLUDGES ON SOIL PHYSICAL PROPERTIES

SESSION 2 - EFFECTS OF SEWAGE SLUDGES ON SOIL BIOLOGICAL PROPERTIES

PREFACE

The agricultural value of sewage sludges is well known and a lot of published data has demonstrated the positive effects of sludge applications on plant growth and yield.

These effects are probably due mainly to the nitrogen and phosphorus content of sewage sludges. But, as sludges are more organic than mineral, we can expect an effect of the organic matter added to the soil on soil fertility.

Certainly, in the future, landspreading of sludges will be regulated, taking into account pollution hazards for waters (excess of nitrogen and phosphorus supply compared to plant needs and soil storage capacities) and for soils (excess of heavy metals supply and build up in soils).

There will be regulations fixing what low level of sludges may be spread each year, decreasing their comparative value with respect to mineral fertilizers.

In this eventuality, the organic value of sludges will take on a greater importance and several questions arise :

- what is the lowest amount of sludge to be spread to have an immediate effect on soil physical properties?

- are sludges effective on soil physical properties when spreading repeated low amounts?

On the other hand, organic matter and soil biology are closely linked and there are few data on the possible effects - beneficial or detrimental - on soil organisms.

The purpose of the Seminar, held in Munich, 23 - 24 June 1981, organized by Dr Süss from Bayerische Landesanstalt für Bodenkultur und Pflanzenbau, Munich, and Dr. Catroux from INRA, Dijon, and sponsored by the Commission of the European Communities, was to collect data, and exchange ideas on these two important practical problems related to the agronomic value of sewage sludges.

SESSION 1

EFFECTS OF SEWAGE SLUDGES ON SOIL PHYSICAL PROPERTIES

Increasing organic matter of soils by sewage sludge

The analysis of nitrogen and humus in connection with a field
test for the fertilization with straw and sewage sludge

Influence of limed sludge on soil organic matter and soil
physical properties

Some effects of sewage sludge on soil physical conditions and
plant growth

General comments on the organic value of sludge

Influence of sewage sludge application on physical properties of
soils and its contribution to the humus balance

Influence of increasing amounts of sewage sludge on the soil
structure

The influence of the agricultural utilization of domestic sewage
sludge on the quality of the soil

Physical properties in sewage sludge and sludge treated soils

Modifications of some physical and chemical soil properties
following sludge and compost applications

INCREASING ORGANIC MATTER OF SOILS BY SEWAGE SLUDGE

by

TH. DIEZ

1. INTRODUCTION

Sewage sludge from Bavarian municipal sewage plants with mechanical and biological purification contains about 35 to 55 %, in average 46 % organic matter, according to our analyses of 1977-1979.

Sewage sludge therefore can generally be used for improving the organic matter supply of soils.

In two field trials with different soils the effect of continued application of sewage sludge (ca. 8 t dry matter/ha per annum) on the organic matter content of the soils was studied.

In addition the shallow gravel soils in the north of Munich were analyzed, which had received extremely high amounts of sewage sludge (19 respectively 14.5 t dry matter per ha and year) over several decades. Changes in soil profile due to sludge application were tried to be evaluated.

Moreover, investigations were made, studying the effect of these extremely high amounts of sludge on microbiology of the soils treated.

2. METHODICAL APPROACH

Calculating the increase of organic matter, the humus contents of sludge treated and non-treated lots were determined and the difference between both was compared with the amount of organic matter added by sludge. Bulk density of the soils was determined by means of 1000 cm^3 metal cylinders with all well known difficulties of analysing gravelly soils.

It had to be taken into account that the increase of organic matter had changed the bulk density of the fine textured soil considerably.

Evaluating the input of organic matter added by sludge a mean value of 46 % related to dry solids was chosen.

The obtained rates of decay can only be approximate values because of the uncertainty of the basic data, e.g. amount of sludge disposal, real organic matter content of the sludge, possible heterogenity of the compared soils, errors of determination of bulk density and content of gravel.

3. RESULTS

3.1. Increase of organic matter

3.1.1. Field_trials_with_sludge,_Puch_experiment_station

The trials were established on two different soils, one derived from calcareous gravel, the other from loess. Climatic conditions include a mean annual precipitation of 890 mm and a mean annual temperature of 7.8°C. Plough depth of both sites was 20 cm.

Considering the amount of stone and the different bulk density the weight of the fine textured top soil is determined as 2520 t for the gravel-soil and 3000 t for the loess-soil (table 1).

The input of 24.5 t organic matter by sludge within 6 years resulted in a considerable increase of the organic matter content in both soils, about 0.34 %, i.e. 8.6 t/ha in the gravel-soil, about 0.43 %, i.e. 12.9 t/ha in the loess-soil. That makes a decay of organic matter of 65 % for the gravel-soil, of 53 % for the loess-soil.

The more intensive humus decay in the gravel soil can be explained with the better aeration of this soil.

3.1.2. Sludge treated lots of the city of Munich

Investigations were made on

- a lot untreated (without sewage sludge)
- a lot having received about 400 t dry solids of sludge since 1959
- a lot that has been sludge-treated since the opening of the sewage plant in 1925, having received approximately 800 t dry solids in the meantime.

The three soils differ especially in the depth of solum and the organic matter content (see table 2).

The original soil, still preserved in a heath area, has a solum of about 18 cm. In the course of the very intensive sludge treatment the soils were ploughed deeper and deeper. Plough depth for the soil with the longest time of treatment today is about double as deep as before sludge application.

As a consequence of the sludge treatment the humus percentage increased considerably and attains values close to peaty soils for the most intensively treated soil. The bulk density drops correspondingly.

For both sludge treated soils the decay of the organic matter added by sludge attains similar values of 41 and 45 % respectively. This rate of decay is about one third lower than that of the very similar gravel soil of Puch.

Table 1

Increase of organic matter in two soils derived from limeston gravel and loess respectively by hygh amounts of sewage sludge (ss); Experimentstation Puch, Lkrs. Fürstenfeldbruck
Period : 1983-1978

	Gravel soil	Loess soil
Soil profile	Ap 0-20 C-Gravel	Ap 0-20 B_v 20-50+
Stones, vol. %	10	-
Top soil without stones, CM	18	20
x bulk density	1.4	1.5
x 100		
= Top soil without stones, T/HA	2520	3000
Organic matter without sludge, %	3.44	2.23
Organic matter with sludge application, %	3.78	2.66
Increase of org. matter; T/HA	8.6	12.9
Org. matter added by sludge within 6 years, T/HA	24.5	24.5
Org. matter decay in % of org. matter added by sludge	65	53

Table 2

Increase of organic matter in soils of limestone gravel north of Munich due to extremly high application of sewage sludge (ss)

	Without ss	400 T ss (dry matter) since 1959	800 T 1925
Soil profile	A_h 0-8 BC 8-18 C-Gravel	A_p 0-28 C-Gravel	A_p 0-38 C-Gravel
Solum, CM	18	28	38
- Stones, Vol. %	38	37	35
= Fine textured soil, CM	11.2	17.7	24.7
x bulk density	1.24	0.96	0.76
x 100			
= Fine textured soil, T/HA	1389	1700	1877
Organic matter, %	5.4	10.8	14.8
Organic matter, T/HA	75	184	278
Increase of org. matter, T/HA	-	109	203
Organic matter added by sewage sludge, T/HA		(9 t/year)	(6.7 t/year)
Organic matter decay in % of organic matter added	-	41	45

There are 3 reasons for explication :

1. While the soil at Puch could digest the relatively small
   amounts of sludge (maximum 10-20 t dry solids/ha in one
   application), it is probable that the high amounts of sludge
   (up to 100 t in one) added to the Munich soils, have disturbed microbiological activities at least temporarily.

2. The Munich soils are very deeply ploughed. Organic matter
   brought into the badly aerated deep zones has undergone less
   decay.

3. The very high amounts of sludge applied within the last years
   were stapled in deponies for 1-2 years before they were
   brought to the fields. During the intermediate deposit a
   first decay of the easily decayable parts took place, i.e.
   the finally added sludge was more stabilized.

3.2. Biological activity

Decay of organic matter is a cumulative effect of biological
activity in the past. Additional investigations should show the
effects of the very intensive sludge application to the Munich
soils on the actual parameters of biological activity.

Therefore the above described soils and - in addition - a relatively little treated soil were analysed by determining mineralization of nitrogen, saccharase, proteinase, biomass and
number of soil bacteria (1) (see graph).

Results can be summarized as follows : Biological activity being
very high already in the untreated soil (pasture land) has been
increased furthermore by adding sewage sludge.

Up to a sludge application of 400 t/ha the various biological
activities follow the increasing content of organic matter (C_t)
more or less. At a higher sludge level the increase rates of
biological activities become smaller or even negative. May be

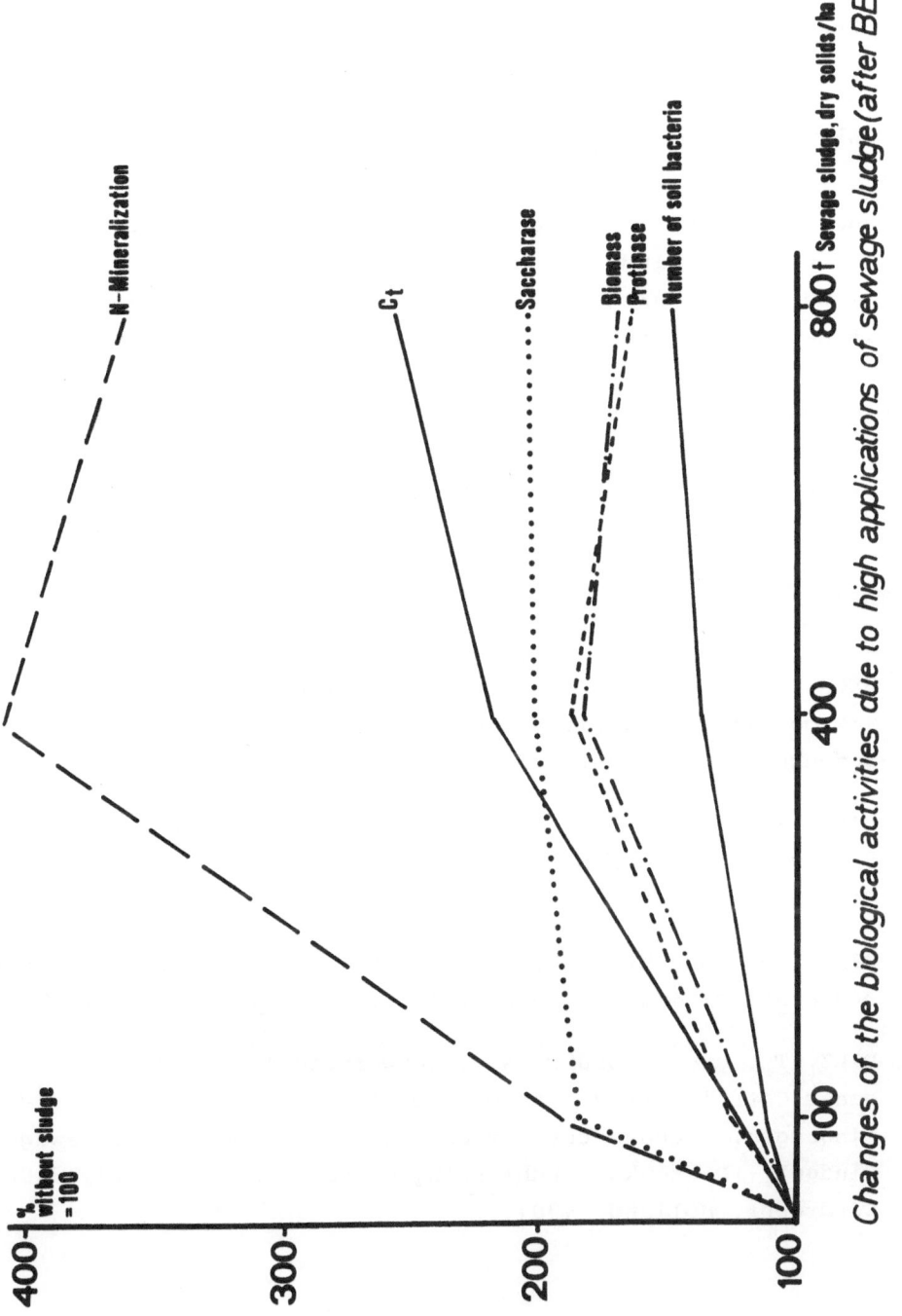

Changes of the biological activities due to high applications of sewage sludge (after BECK)

this is the beginning of toxic effects of the very strong
heavy metal contamination of these soils. The subject was
discussed at the last meeting of Cost 68 in Vienna in 1980 (4).

SUMMARY

The decay of organic matter, continuously added to different
soils by sewage sludge attains values between 41 and 65 %.
The lower values are valid for less aerated and deeply ploughed
soils and soils with high amounts of sludge application.

Biological activity seems to be favoured by sewage sludge.
Only at an extremely high level it shows a slight drop possibly
due to heavy metal contamination of the soil.

LITERATUR

1. BECK, Th. und SÜß, A. : Der Einfluß von Klärschlamm auf die
   mikrobielle Tätigkeit im Boden. - Z. Pflanzenernährung und
   Bodenkunde, 142, 299, 1979.

2. DIEZ, Th. und ROSOPULO, A. : Schwermetallgehalte in Böden und
   Pflanzen nach extrem hohen Klärschlammgaben. - Landwirtsch.
   Forsch. 33/I. Sh., 236-248, 1977.

3. DIEZ, Th., und WEIGELT, H. : Zur Düngewirkung von Müllkompost
   und Klärschlamm. - Landwirtsch. Forsch. 33, 1, 47-66, 1980.

4. DIEZ, Th. und ROSOPULO, A. : Schwermetallaufnahme verschiedener Getreidearten aus hochbelasteten Böden unter Feldbe-
   dingungen. - Characterization, Treatment and Use of Sewage
   Sludge, - D. Reidel Publishing Company, P.O. Box 17, 3300 AA
   Dordrecht, Holland, 1981.

DISCUSSION

Dr WILLIAMS   : What particular metals and at what levels were biological activities affected?

Dr DIEZ   : The field trials do not allow to specify effects for certain metals.
Heavy metal contents in the soil after application of 800 t sludge reach values of Cu 340, Zn 1860, Cr 175, Ni 50, Pb 1340, Cd 42 ppm.

Dr DE HAAN   : Were the sludges used in your experiments of the same type (wet sludges on the one hand and dried sludges on the other hand)?
Was there decomposition of the dried sludges during the drying process?

Dr DIEZ   : The sludge treatment of the soils investigated changed during the application period of 20-50 years. In the early years wet sludge was commonly used, in the last years sludge of about 20% dry solids, dewatered during a 1-year deposition in pounds.

Dr DANNEBERG   : Did you observe any phytotoxic effect due to heavy metals?

Dr DIEZ   : Not at all.

Dr DANNEBERG   : Are the increases in organic matter statistically proof?

Dr DIEZ   : No

Dr CATROUX   : Which kind of Biomass and Ammonification method did you use?

Dr BECK   : For the determination of total microbial
                     biomass; including bacteria and fungi, we
                     used the new physiological method of Anderson
                     and Domsch (Soil Biol. Biochem. 10, 215-221)
                     based on short-term measurement of respiration
                     after saturation of soil with glucose.
                     Ammonification was measured after a 20-days
                     incubation of fresh soils under constant
                     laboratory condition by determination of
                     NH_4-N and NO_3-N mineralized from soil organic
                     matter.

Dr CATROUX   : Did you observe any difference in soil
                     "workability" on your field trials?

Dr DIEZ   : Yes. The highly sludge treated soils are some
                     what more clumpy and tend to stick to the
                     working tools.

Dr SUESS   : Concerning the organic matter increase in
                     your field trials, did you try to make a
                     difference between O.M. coming from sludges
                     and O.M. coming from plant residues?

Dr DIEZ   : No, this was not possible. But O.M. from
                     plant residues would certainly not have been
                     enough to explain any increase of organic
                     matter in the soils.

THE ANALYSIS OF NITROGEN AND HUMUS IN CONNECTION WITH A FIELD

TEST FOR THE FERTILIZATION WITH STRAW AND SEWAGE SLUDGE

O.H. DANNEBERG, G. STORCHSCHNABEL AND S.M. ULLAH

1. INTRODUCTION

Recycling of sewage sludge on agricultural lands seems an inevitable part of waste disposal. It is reported to 1.) be a good fertilizer (Lunt, 1959; Hinesly et al., 1972 and Sommer, 1977), 2.) improve soil aggregation (Epstein, 1975), 3.) act as soil conditioner both physically and chemically (Hinesly et al. 1972; Epstein, 1975; Epstein et al., 1976 and Ming-hung Wong, 1979) and 4.) provide needed organic matter to the soil.

Despite its beneficial effects, the random application of sewage sludge in agriculture is restricted because of actual or potential hazards due to the presence of heavy metals, organic toxicants and disease producing organisms. When industrial effluents are discharged, sludge can contain excessive amounts of heavy metals and organic toxins. Domestic sludge, on the other hand, would expectedly be low in such contaminants (Lee et al., 1980).

In agriculture, sewage sludge with an average of 5 % dry matter can be utilized using application rates of 50 to 100 m^3/ha in 2-3 years cycle (v. Boguslawski, 1980). If application rates are too high, excess NO_3-N accumulates in the soil (King et al., 1972 and Kelling et al., 1977) and may be taken up by crops in excess of needs or reach the ground water.

Both pot and field experiments proved the efficiency of sewage sludge in enhancing growth and yield of crops (Cunningham et al., 1975; Singh et al., 1975; Haunold et al., 1977; Kelling et al.,

1977; Sheaffer et al., 1979; Zwarich et al., 1979; Lembke et al., 1980 and Magdoff et al., 1980) due to higher uptake of N, P, Ca and Mg (Sheaffer et al., 1979); but some depression in yields was also reported at higher rates possibly due to soluble salts and phytotoxic concentrations of heavy metals (Cunningham et al., 1975 and Kelling et al., 1977). However, a new sludge fertilization technique in combination with straw has been introduced because it supplies adequate water and necessary N for the decomposition of straw. Nitrogen, thereby, being immobilized is prevented from leaching loss. This immobilized N becomes again available to the plants after mineralization (Danneberg et al., 1980).

In a series of pot experiments, the promising effect of straw-sewage sludge combination was observed (Haunold et al., 1977 and Danneberg et al., 1980). Now this has been tested in field experiments under natural condition in the Austrian main production region, Marchfeld, to prove their efficiency with regard to N_{min}, soil humus and ultimate bearing on yields of crops. The results of two years (1978/79 and 1979/80) are discussed.

2. MATERIALS AND METHODS

The field trial was conducted at the experimental station of Großenzersdorf of the Institute of Plant Production and Plant Breeding of the Agricultural University of Vienna, Austria. The experimental field having an area of 100 m x 135 m was divided into 3 main plots (with 35, 30 and 35 m width), each of which was further subdivided into 11 subplots with a width of 8 m. These subplots were separated from each other by a strip of 0.4 m.
Treatments are as follows :

1. Mineral fertilization. (standard, S)
2. Control (C)
3. Straw (St)
4. Straw + water (St + W)

5. Straw + Sewage sludge (St + Sl)
6. Sewage sludge (Sl)

In order to eliminate any difference among the subplots, each
of the treatments (2-6) was inserted between 2 standard plots
(treat. 1; Boguslawski et al., 1972).

Winter wheat (WW), winter rye (WR) and spring barley (SB) were
the test plants, which are rotated for three years – once in
each of the three main plots. Rye was sown after ploughing in
autumn, wheat and barley in October and March respectively.

Chopped winter wheat straw was applied to the plots with treatments 3, 4 and 5 at the rate of 5000 kg/ha. P and K as DC-45
were applied in all plots at the rate of 700 kg/ha in August.
Sewage sludge was spread at a rate of 200 m^3/ha on plots with
treatments 5 and 6 at the end of August. Treatment 4 received
the same quantity of water.

For the first year (1978/79), 90 kg N/ha was applied to the
standard plots in two rates during the vegetative growth of
crops. Sludge treated plots (treat. 5 and 6) received only
45 kg N/ha, whereas the straw treated plots (treat. 3 and 4)
received 140 kg N/ha in three rates – 50 kg after straw application at the end of August and the rest 90 kg along with the
standard plots in two rates.

In the second year, 1979/80, plots under winter wheat and
spring barley with treatments 1, 3 and 4 received 80 kg N/ha
in two rates during their growing periods – 30 kg at the beginning of growth and 50 kg at the tillering stage. Additionally
treatments 3 and 4 received 50 kg N/ha in August in order to
prevent any possible N-deficiency resulting from straw decomposition. No nitrogenous fertilizer was applied to the plots
treated with sewage sludge (treat. 5 and 6) in this year.

Treatments 1, 3 and 4 under WR received 30 kg N/ha in the beginning of April. Further 50 kg N/ha were added to the straw
treated plots (treat. 3 and 4) at the end of August.

Soil samples were collected every three weeks from depths of
0-25, 25-60 and 60-90 cm starting from the beginning of growth
to the maturity of crops. The soil samples were transported to
the laboratory in insulated ice-boxes and were immediately analysed for N_{min} contents (Scharpf, 1977). The soil samples were
extracted with 2N KCl. The N_{min} was determined by steam distil-
lation with MgO and Devardas' alloy (Bremmer, 1965). Humic
substances were extracted from air-dried soil samples (0-25)
with a chelating resin and water. Extracts were fractionated
by precipitation into fulvic acids, brown humic acids and grey
humic acids. Optical densities of these fractions were measured
at 400 nm (Danneberg and Schaffer, 1974).

The crops were harvested with eight replications in 1979 and
with 12 replications in 1980 from an area of 1.25 x 8 m^2 for
each replication. The yield/ha was determined.

3. RESULTS AND DISCUSSIONS

3.1. N_{min} Content :

The mineral N available to the crops in the rootpenetrated zones
of the soil profile was between 90-150 kg/ha at the early growing season and 50-80 kg/ha at maturity in the first year (1978/
79). In the second year (1979/80), it ranged from 140-179 kg/ha
at the beginning of growth and 59-94 kg/ha at maturity. This
available N resulted from organic and inorganic fertilization
and through mineralization of soil organic matter. The decrease
in N_{min} content with time was due to plant uptake and possible
losses of N. The decrease in N_{min} varied with the crops. The
decrease in N_{min} was noticed in all three crops for both years,
except under WR for the first year (Fig. 1 and 2). Possibly the
time interval of sampling under WR in the second year was too
large to follow the change in N_{min} content sufficiently.

In the first year, the treatments 3-6 did not raise N_{min} significantly compared to the standard (treat. 1). However, sludge

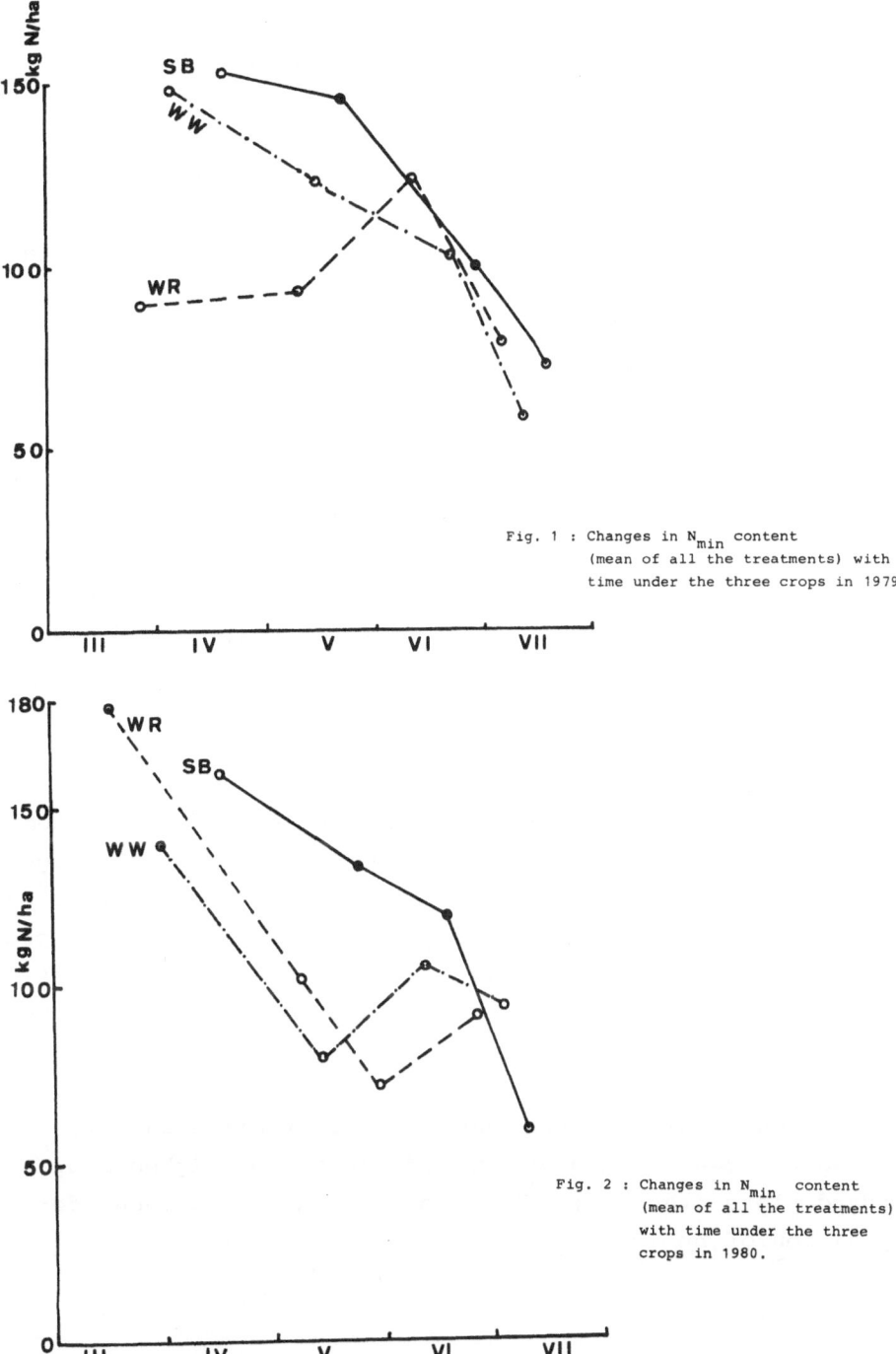

Fig. 1 : Changes in N_{min} content
(mean of all the treatments) with
time under the three crops in 1979.

Fig. 2 : Changes in N_{min} content
(mean of all the treatments)
with time under the three
crops in 1980.

treatments (5 and 6) differed significantly from the control
(treat. 2) under all crops, except treatment 6 under barley
(table 1).

Mineral-N in treatment 6 under barley was significantly lower
than the other treatments (3-5) possibly because of its losses
through leaching during winter time and volatilization in May
as a result of lacking in surface cover and drought

In the second year, the N_{min} content was significantly higher
at sludge amendment (treat. 6) than the standard (treat. 1) in
case of wheat and barley (table 1). Just as in the previous
year, no difference was observed between the standard and other
treatments (3-5). The control was all along significantly much
lower than the standard as well as all other treatments (3-6;
table 1). No difference was observed between straw treatments
(3 and 4) and also between sludge amendments (treat. 5 and 6).

3.2. Humic substances

In the first year, a significant increase in extractable humic
substances was observed as a result of application of sewage
sludge alone and in combination with straw (table 2). This increase was noticed especially in grey humic acid fraction probably because of synthesis and subsequent polymerization of
humic acids. A similar increase was also reported in pot experiments (Danneberg et al., 1978).

In the second year, however, no significant difference was
observed between the sludge treated plots and neighbouring
standard plots (table 2). This might be due to enhanced decomposition of humic substances.

TABLE 1 : NITROGEN CONTENT IN KG N/HA (0-90 cm) (DIFFERENCE TO STANDARD)

Treatment	WR		WW		SB	
	1979	1980	1979	1980	1979	1980
1 S	0	0	0	0	0	0
2 C	-13,25	-11,25	-7,33	-15,75	-10,50	-17,00
3 St	- 3,58	6,83	3,75	- 1,83	5,83	- 5,75
4 St + W	- 5,92	6,08	-2,25	2,25	3,83	- 8,50
5 St + Sl	0,50	2,58	11,25	4,00	5,33	5,33
6 Sl	2,50	2,75	7,42	9,67	- 9,50	13,50
LSD 5 %	13,53	10,24	11,69	8,95	10,41	8,84
Average N content of standard	90	109	105	106	113	120

TABLE 2 : CONTENT OF HUMIC SUBSTANCES IN OD/G SOIL (0-25 cm) (DIFFERENCE TO STANDARD)

Treatment	WR		WW		SB	
	1979	1980	1979	1980	1979	1980
1 S	0	0	0	0	0	0
2 C	0,28	-1,48	-0,79	1,37	0,59	0,03
3 St	-0,61	1,41	2,53	0,26	1,56	0,12
4 St + W	3,82	-0,17	3,80	-0,25	2,66	0,87
5 St + Sl	7,17	0,59	5,17	0,67	3,39	0,67
6 Sl	5,78	0,93	4,38	0,93	5,85	0,77
LSD 5 %	1,53	1,88	1,17	1,01	1,14	1,08

TABLE 3 : YIELD DATA IN KG/HA (DIFFERENCE TO STANDARD)

Treatment	WR		WW		SB	
	1979	1980	1979	1980	1979	1980
1 S	0	0	0	0	0	0
2 C	-535	-636	-757	-1035	-124	-1976
3 St	405	- 73	2	452	-273	308
4 St + W	30	- 19	136	299	-255	203
5 St + Sl	- 4	-380	244	- 60	288	-138
6 Sl	-156	-128	76	- 52	3	31
LSD 5 %	458	503	247	555	396	343
Average standard yield	3037	3511	3469	4881	4053	5203

3.3. Yields

In the first year, the control was significantly lower than the standard, except under barley (table 3). Other treatments (3-6) did not differ from the standard, which showed that the treatments 3-6 were capable of raising the yield to the same extent as the mineral nitrogen used in the standard treatment. Sludge amendments (treat. 5 and 6) depressed yields of rye to some extent compared to the standard (table 3). This might be due to lodging in the beginning of May 1979, because of excessive vegetative growth probably due to higher uptake of nitrogen. Kelling et al. (1977) also reported some depression in yield at higher rates of sludge.

In the second year, significantly lower yields were also noticed at control condition under all crops (table 3). Like in the previous year, there was no significant difference in yields between treatments 3-6 and the standard. In this year, some, but not significant, decrease in yield under rye in sludge treated plots (treat. 5 and 6) was observed because of lodging due to excessive vegetative growth.

Similar increase in yields from sludge application to soil have been reported by many authors - in corn (Magdoff et al., 1980; Lembke et al., 1980; Cunningham et al., 1975; Kelling et al., 1977; Sheaffer et al., 1979; Singh et al., 1975), wheat (Sabey et al., 1977), vegetables (Lunt, 1959), and soybeans (Hinesly et al., 1976). Singh et al. (1975) noted that an application of 30 tons of sludge per acre more than doubled the corn yield compared to the check plots. The greater response to sludge than N fertilizers was probably caused by other factors than N contribution (Magdoff et al., 1980) and attributed to better soil structure (Epstein et al., 1976) resulting from sludge application and increased water holding capacity (Sawhney et al., 1980).

The yields in the first year were much lower than that of the second year under all crops because of drought in the first year, which affected yields severely (table 3).

4. CONCLUSION

Sludge application, alone and in combination with straw, raised soil-N content and consequently crop yield to the same extent as mineral nitrogen fertilization. From the second year, no addition of mineral nitrogen in sludge treated plots was necessary to obtain optimum yield. Elevated N levels were sufficient for wheat and barley, but somewhat too high for rye, which showed some lodging in early summer.

5. SUMMARY

A field test under the conditions of the Austrian Marchfeld is carried out to investigate the practical use of a fertilization with sewage sludge and/or straw. The test comprises : control (without nitrogen), mineral fertilization, straw fertilization, irrigated straw fertilization, straw-sewage fertilization and sewage sludge fertilization. A full-rotation of winter wheat, winter rye and of spring barley is used as test plantation.

Samples, taken continuously from the plots at the time of vegetation, were investigated with respect to the contents of mineral nitrogen useful for plants in the rooted soil. A photometric humus-complex-analysis is used to investigate the soil samples of the uppermost 25 cm-layer with respect to their contents of extractable humates.

Application of sewage sludge alone and in combination with straw raised the levels of soil nitrogen as well as yield to the same extent as the standard, where fertilizer nitrogen was used. Both soil nitrogen and crop yield were significantly much higher at sludge treatments than the control (without nitrogen) in both years (1978/79 and 1979/80). This increased N_{min} was found to be adequate for wheat and barley; but somewhat too high for rye, which showed some lodging in May.

A significant increase in extractable humic substances was observed in the first year at sludge treatments due to synthesis and subsequent polymerization of humic substances; but in the second year, the increase in humic substances was not significant.

6. ACKNOWLEDGEMENT

The authors express their sincere thanks to the Ministry of Health and Environment Protection of Austria for the financial support in order to carry out this work.

7. REFERENCES

v. BOGUSLAWSKI, E., 1980 : Z. Acker- und Pflanzenbau 149 : 406-423.

v. BOGUSLAWSKI, E. and W. SCHUSTER, 1972: in: K.Scharrer und Linser, Hsg.: Handbuch der Pflanzenernährung und Düngung. Springer, Wien u. New York, p. 1181.

BREMMER, J.M., 1965 : in : C.A. Black, ed. : Methods of Soil Analysis. Agronomy 9, Am. Soc. Agron. Madison, 1179-1237.

CUNNINGHAM, J.D., D.R. KEENEY and J.A. RYAN, 1975 : J. Environ. Qual. 4: 448-454.

DANNEBERG, O.H. and K. SCHAFFER, 1974: Bodenkultur 25 : 360-368.

DANNEBERG, O.H. and H. SISTANI, 1978 : Z. Pflanzenernähr. Bodenkunde 141 : 353-365.

DANNEBERG, O.H., E. HAUNOLD, O. HORAK and J. ZVARA, 1980 : Wiener Mitt. Wasser, Abwasser, Gewässer 34.

EPSTEIN, E., 1975 : J. Environ. Qual. 4 : 139-142.

EPSTEIN, E., J.M. TAYLOR and R.L. CHANEY, 1976: J. Environ. Qual. 5 : 422-426.

HAUNOLD, E. and J. ZVARA, 1977: Bodenkultur 28: 270-276.

HINESLY, T.D., R.L. JONES and E.L. ZIEGLER, 1972: Compost Sci. 13 : 26-30.

HINESLY, T.D., R.L. JONES, J.J. TYLER and E.L. ZIEGLER, 1976: J. Water Pollut. Cont. Fed. 48: 2137-2152.

KELLING, K.A., L.M. WALSH, D.R. KEENEY, J.A. RYAN and A.E. PETERSON, 1977 : J. Environ. Qual. 6 : 345-352.

KING, L.D. and H.D. MORIS, 1972, J. Environ. Qual. 1 : 442-446.

LEE, C.Y., W.F. SHIPE, jr., L.M. MAYLOR, C.A. BACHE, P.C. WSZOLEK, W.H. GUTENMANN and D.J. LISK, 1980: Nutrition Reports International 21 : 733-738.

LEMBKE, W.D. and M.D. THORNE, 1980 : Transactions of the ASAE published by American Soc. of Agricul. Eng. 23 : 1153-1156.

LUNT, H.A., 1959 : Connecticut (New Haven) Agric. Stn. Bull. No. 622.

MAGDOFF, F.R. and J.F. AMADON, 1980 : J. Environ. Qual 9 : 451-455.

SABEY, B.R., N.N. AGBIN and D.C. MARKSTORM, 1977: J. Environ. Qual. 6 : 52-58.

SAWHNEY, B.L. and W.A. NORVELL, 1980: The Connecticut Agric. Expt. Stn. (New Haven), Bull. 788 : 1-22.

SCHARPF, H.C., 1977: Dissert. an der Techn. Univ. Hannover.

SHEAFFER, C.C., A.M. DECKER, R.L. CHANEY and L.W. DOUGLASS, 1979 : J. Environ. Qual. 8 : 450-459.

SINGH, R.M., R.F. KEEFER and D.H. HORVATH, 1975: Compost Sci.- J. of Waste Recycling, 16, No. 2.

SOMMER, L.E., 1977 : J. Environ. Qual. 5: 303-306.

WONG, M.H., 1979 : Bull. Environ. Contam. Toxicol. 23: 717-724.

ZWARICH, M.A. and J.G. MILLS, 1979: Can. J. Soil Sci. 59 : 231-239.

DISCUSSION

Dr. WILLIAMS : Could the quantity of straw ploughed in be responsible for immobilizing soil nitrogen resulting in an inadequate supply for the crop?

Dr. DANNEBERG : No, the supply of nitrogen was sufficient in all treatment except the control. In the sludge treated plot the N-supply was always higher than necessary for optimal growth.

Dr. WILLIAMS : In connection with nitrogen immobilization, what quantity of straw was mixed into the soil each year ?

Dr. DANNEBERG : 5 t/ha of wheat straw was ploughed in each year equivalent to production of straw from our average wheat crop.

Dr. DE HAAN : Excess of N applied with the sludges may have been the reason for the poor effect of your sludges in the first year (too much vegetative growth at the cost of grain production ?)

Dr. DANNEBERG : We were, in fact, somewhat high in nitrogen.

Dr. DIEZ : The no-effect of sludge in your field trials may depend on the high quality of the soil you made your experiments (chernozem). On bad soils you get better sludge-effects.

Dr. CATROUX : Did you observe a nitrogen immobilization in the straw-sludge treatment ?

Dr. DANNEBERG : We did not measure it but we are sure that some immobilization occured.

Dr. DE HAAN : Was the sum of fulvic acid + brown humid acid + grey humid acids equal to total organic matter ?

Dr. DANNEBERG : No, it is less than total organic matter. It corresponds to the sum of extractable humic substances.

Dr. MOREL : A comment.
We have made some organic extractions on sludge using NaOH. I agree with you on the fact that sludges are particularly rich in fulvic acid fraction as compared to soils. They presented AF/AH ratio higher than 3. However, it should be mentionned that the rate of extraction was rather low, less than 5 %.

Dr. SÜSS : The methods for analysis of organic matter in soil are different and the results are different. One way to get comparable results is to use the same method all the time. We are looking the different humus fractions as fulvic acids, brown and grey humus acids.

Dr. DE HAAN : A comment.
We should in the first place determine total organic matter in soils. Let us hope we do that all in the same way within the European Community and that no differences arise from using different methods for humus determination. That should be checked.
As to a subdivision of humus in different groups with different values for soil productivity or fertilizy. I quote a pronouncement of Dr. SCHNIZER (Canada) who in a congress about soil organic matter in Brunswick some years ago said that even with the most sophisticated methods, it would take still at least 200 years before the nature of such a complex material as soil organic matter would have been elucidated in a sufficient way for practical purposes.

Dr. DANNEBERG : Endeed it was Prof. SCHNITZER who encouraged the development and use of new and more sophisticated methods in soil organic research to obtain further progess.

Dr. CHAUSSOD :   You recorded the soil mineral nitrogen content during the time of vegetation. But did you record also the nitrogen content of the crop at the same time ?

Dr. DANNEBERG :   No, we did not.

INFLUENCE OF LIMED SLUDGE ON SOIL ORGANIC MATTER AND SOIL

PHYSICAL PROPERTIES

J.L. MOREL, A. GUCKERT

1. INTRODUCTION

Some of the methods used for sewage sludge treatment may greatly influence the agronomic properties of the resulting product. One of these is liming which is employed to stabilize and flocculate sludge. With the high rates of lime actualy used in the treatment plants (10 to 50 % of the dry matter (1), it is obvious that sludge quality should be widely modify. Beyond an important dilution of the sludge components (1), liming leads to the loss of nitrogen by NH_3 volatilisation (2), and when applied on crop land limed sludges produce a reduction not irreversible, however, of the soil phosphate mobility (3-4). Among the positive effects it should be reminded that liming is followed by the reduction of 98 to 100 % of the pathogenic germs present in the sludge (5), and that no detrimental effects on cultivated plants were observed after limed sludge applications even at high loading rates but increasing of production yields were obtained (6-7). In addition to this two essentiel aspects of limed sludge agronomic properties will be developed in this paper : its organic value and especially its role on some soil physical characteristics. Some of the results reported here after have been already published concerning limed sludge organic matter behaviour in soil (8), and sludge consequences on soil structure and water retention (9). All were obtained from the experiments described below.

2. MATERIAL AND METHODS

Two types of experiments were conducted in order to study the limed sludge agronomic properties :

- Laboratory incubation studies were initiated to determine the respiration of sludge amended soils. These, previously described (10), consist in the measure of CO_2 evolved from soil and soil-sludge mixtures incubated at 28°C. The soil was selected for its poor structure stability. It was a loamy clay soil presenting a 7,2 pH, and medium levels of C, N,P.

- A long term field experiment was set up in 1974 on the experimental fields of La Bouzule (Meurthe-et-Moselle, France), on a loamy clay soil (typic brown leached soil) it has been described in detail before (8-9). Four applications of limed sludge + Fe Cl_3 issued from the Nancy-Maxeville treatment plant were made. Among the treatments, two loading rates are reported in this paper.

A summary of the treatments is presented in figure 1. Each treatment was replicated four times. Results reported here concern soil samples collected from the Ap horizon of each plot 2,5 and 5,5 years after the first sludge application. Thus, in 1979, it was possible to measure a four years residual effects of limed sludges because only half of the limed sludge treated plots received a fourth application (see figure).

Soils issued from these experiments were analysed for :

- organic matter : total organic carbon was determined by combustion using the Carmhograph 12 of Wösthoff apparatus.

- structure stability : soil aggregate stability was measured according to Henin and al. 1958. Soil water stable aggregates were determined after two types of treatments, ethanol and benzene, and without treatment, in order to appreciate respectively soil cohesion, soil wettability and direct effects of water on soil stability.

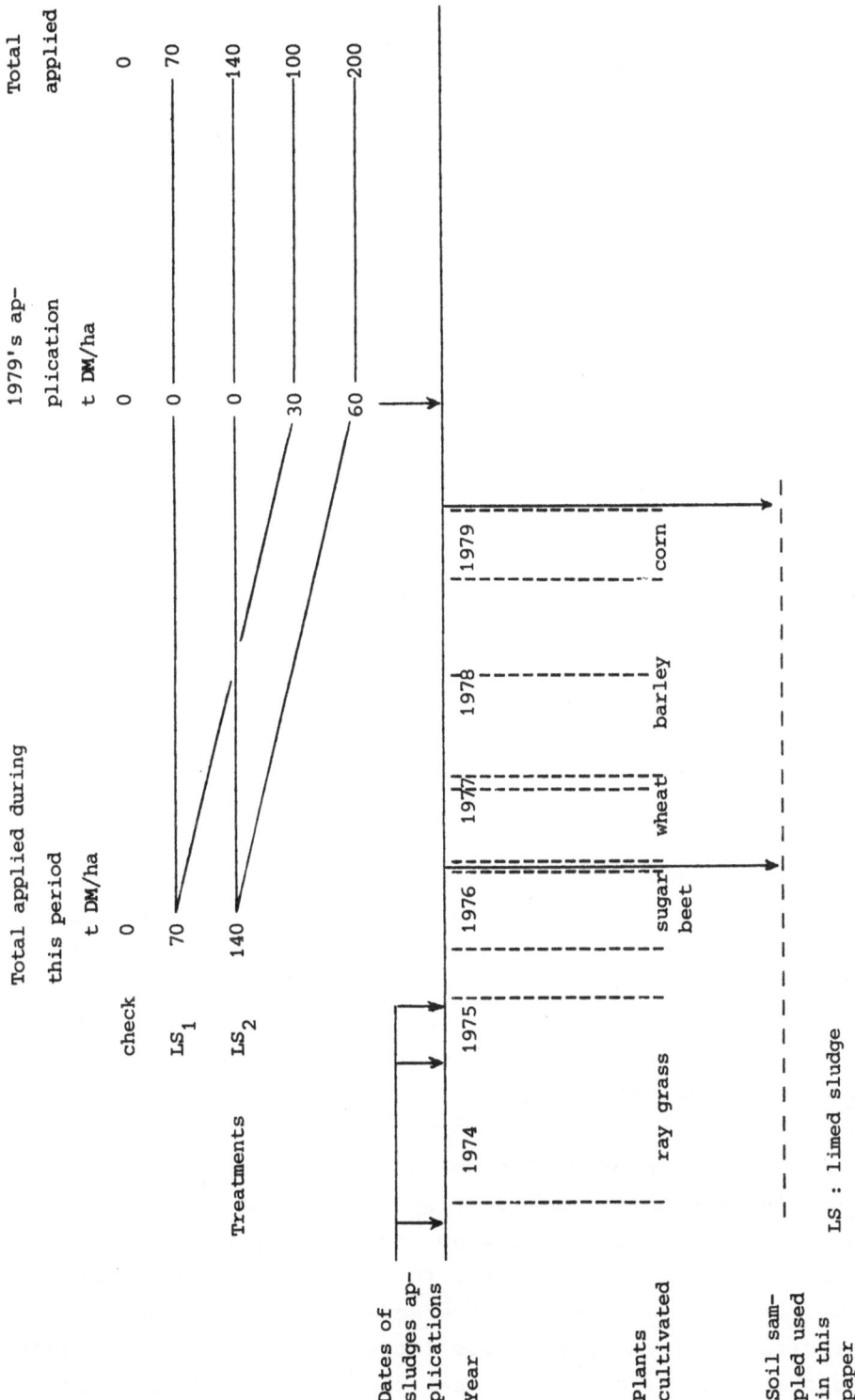

Figure 1 : Summary of the field experiment

- hydraulic conductivity : it was determined according to Henin
  and al. 1958. An hydraulic conductivity coefficient was measured by percolation of water through a soil column.

- water retention : soil water retention was measured at two
  potential values : pH 2,5 and pF 4,2 corresponding to the
  field capacity and to the wilting point. The soil water content between these two values is termed available water. The
  Bouyoucos method and a pressure plate apparatus were used.

3. ORGANIC VALUE OF LIMED SLUDGES

An incubation study of soil-sludge mixtures showed (table 1)
that limed sludge additions produced at the beginning of the
experiment a lower CO_2 evolution than an unlimed sludge applied
at the same organic carbon rate did. However, after the first
thirty days, the respiration was higher in the soil enriched
with limed sludge until the end of the incubation. Then, cumulative results were similar for the two types of soil-sludge
mixtures. The rythm of respiration was rather different : more
uniform with the limed sludge than with the unlimed sludge.
Thus, assuming that the CO_2 evolved from the mixtures represents
the organic matter decomposition, it could be concluded that
liming causes a more regular mineralization of sludge. However,
two factors which could not be taken in account above must not
be omitted : possible under or overestimations of CO_2 evolved
due to the "priming effect" and to the lime carbonatation or
$CaCO_3$ decay during the incubation period.

The carbon analysis of soil samples issued from the field experiment in 1976 shows (table 2) an increase of the soil organic
matter (S.O.M.) content following the sludge applications. The
increments of S.O.M. are related to the sludge loading rates.
However, the S.O.M. content decreased with time as shown by
the results obtained three years later (1979). These decreases
were gratest in the high treatment plots; losses of organic

TABLE 1 : CO_2 EVOLVED FROM SOIL-SLUDGE MIXTURES DURING A SIX MONTHS INCUBATION

Days of incubation	1	7	14	21	30	62	90	125	182	Total C evolved
			mg CO_2 evolved daily							
Check soil	288*	1.68	0.78	0.96	1.50	0.75	0.67	0.55	0.50	91.1
Soil + aerobic sludge	11.10	4.14	1.11	1.18	1.95	0.77	0.76	0.54	0.36	106.6
Soil + anareobic limed + Fe Cl3 sludge	9.42	2.58	0.99	1.14	1.63	0.85	0.95	0.68	0.54	106.3

* mean of three replications.

matter during that period were approximately 11 % in the check
plots while they reached nearly 40 % in the limed sludge
treated plots.

Nevertheless, the organic carbon content in the sludge treated
soils was still greater than in the check soils.

The characterization of the S.O.M. in 1976 showed a high difference between check soils and limed sludge treated soils (8).
The later was characterized by a high proportion of fluvic
acids versus humic acids and a large content of the unextracti-
ble fraction termed humin.

Although, the organic value of sludge has not been clearly
established, it could be assumed that when it is lime treated,
sludge could enhance significatively the S.O.M. content. As a
part of the limed sludge organic matter appears easily decomposable in the soil, a non neglectible fraction shows a
better resistance to decay. The origin of that resistance is
thought to be due to the high content of lime which act at two
levels. Increasing the sludge pH localy reducing the microbial
activity. Changing the organic matter to a status similar to
that of calcareous soils (8). An addition of 20 % of $Ca(OH)_2$
may indeed influence greatly the sludge organic matter evolution in soil. The mechanical and chemical protection processes
of organic matter known in calcareous soils would occur but at
a lower degree in limed sludges. Further, the presence of a
high concentration of iron issued from the flocculate agent may
induce phenomenon reducing the biodegradation of sludge humic
compounds.

Such described processes would be plaintly effective only during
the first year following the application, because limed sludges
are not immediately well mixed with the soil. After that period
sludge is better diluate in the soil thus diminishing its
resistance properties. An extreme case was observed after a
very high rate (1800 t/ha WM) without mixing the sludge with
the soil (11) : there, a 10 cm "limed sludge horizon" was

TABLE 2 : SOIL ORGANIC CARBON CONTENT AFTER LIMED SLUDGE
APPLICATIONS

Date of sampling	check	limed sludge 1*	limed sludge 2 **
		%	
1976	1.83	2.54	3.58
1979	1.62	1.78	2.16

* Total applied : 70 t DM/ha (three applications 30, 10, 30).
** Total applied :140 t DM/ha (three applications 60, 20, 60).

succeded by a typical "calcic mull" where structure and organic matter evolutions were very similar to those of a A_1 horizon of a rendzina.

We should notice on the other hand, that the incorporation of large quantities of Ca could be followed by a "priming effect" producing a loss of the soil organic matter. Besides, the importance of indirect factors on soil organic matter content should not be forgot : carbon supplied by plant residues (roots, shoots) which production is higher on sludge treated plots in relation to the nitrogen effect of sludge.

4. INFLUENCE OF LIMED SLUDGES ON THE SOIL WATER RETENTION

The results of the soil water retention in 1976 for three treatments are summarized in table 3. The field capacity was significantly modified only with the high loading rate (140 MS/ha in

three applications) while the wilting point showed an increase
with the two sludge treatments. However, the available water was
affected only with the highest treatment and an increase of 5 %
was noted.

The field capacity measured on this treatment three years later
showed a significant decrease. The differencies between the
values of each treatment were then slight.

In the years following sludge applications the soil water retention showed modifications. This was due to the contribution of
the sludge organic matter. This was confirmed on sugar beet
which benefited of a better water supplying on limed sludge
treated soils during period (6). However, as indicated before
(9), the available water can be significantly and longly enhanced
only after large limed sludge applications. Such heavy rates are
commonly used for the greening of destroyed soils, strip mine
spoils, sanitary landfill ... In these cases, more durable
changes would be expected with limed sludges because of the
high resistance to biodegradation of their organic matter when
they are not mixed with soil.

TABLE 3 : WATER RETENTION AT FIELD CAPACITY AND WILTING POINT

FOR SOILS ENRICHED WITH LIMED SLUDGES

		check soil	limed sludge 1*	limed sludge 2**
		water content % by weight		
Field capacity	1976	33.2 (a)***	34.6 (a)	41.9 (a)
	1979	31.4	32.7	33.8
Wilting point	1976	12.7 (a)	13.8 (b)	16.6 (c)
"available water"	1976	20.5	20.8	25.3

*   Total applied : 70 t DM/ha (three applications 30, 10, 30).

**   Total applied :140 t DM/ha (three applications 60, 20, 60).

***   Means on the same line followed by the same letter are not
        significantly different at a 5 % level.

5. INFLUENCE OF THE LIMED SLUDGES ON THE SOIL STRUCTURE

It has been shown that sewage sludges could influence the physical state of soils (12,13) modifying particularly the aggregate stability and the soil permeability. Further, it is well known that soil liming produces a more stable soil structure, increasing especially water movement in the soil. Thus, what one could expect about the contribution of limed sludges on the soil structure ?

Table 4 presents the stable aggregates according to the method of Henin and al. 1958 of a soil incubated during six months with two types of sludges, aerobic and anaerobic limed sludge, applied at the same rate of organic carbon. A more important increase was obtained with the limed sludge. This fact was due to the liming effect of the soil and possibly to a better stabilization by calcium of the polyssaccharides produced during the sludge organic matter decomposition and which are thought to reduce the aggregate wettability (14).

Results issued from the field plots 5,5 years after the beginning of the experiment (table 5) show significant increases of the aggregate stability with the increasing rates of limed sludges. The aggregates determined according Henin and al. 1958 were affected by the successive sludges applications. This means that the soil stability was improved and its wettability reduced. This reduction related to the liming effect and to the polysaccharide production could be due also to the high content of oil and grease carbon in sludge (more than 30 % of the total organic carbon according to Hohla and al. 1978) which are more resistant to the biodegradation.

The long term effects of limed sludges on soil stability were evaluated by applying Henin's tests to the soil sampled four years after the last application (1979) (table 6). Considering each sludge treatment, significant improvements could be yet detected. Nevertheless the stable aggregate differences between the check soil and the sludge treatments were rather lower than above. The lowest rate of application of limed sludge gave in particular no significant differences for water stable aggregates after the benzene soil pretreatment indicating that the sludge organic matter effect had dissipated during the last four years. Thus, the efficiency of the limed sludge is as durable as its organic matter is.

Furthermore, it should be pointed out the role of the successive cultures on the soil aggregation. The biomass production was much higher on sludge soil treatments due firstly to the nitrogen effect of the sludge than on the check soils. This produced more organic residues and was followed by a better root activity which resulted in a fine division of soil particles.

TABLE 4 : INFLUENCE OF TWO TYPES OF SLUDGES ON THE STRUCTURE

STABILITY OF AN INCUBATED SOIL

	aerobic sludge	anaerobic limed sludge + Fe Cl_3
	%	
Water stable* aggregates	30.0 (a)	33.3 (b)

* mean of WSA after pretreatments (ethanol-benzene) and without pretreatment

TABLE 5 : INFLUENCE OF LIMED SLUDGE APPLICATIONS ON THE SOIL

STRUCTURE STABILITY (1979)

Water stable aggregates**** %	Check soil	Limed sludge 1*	Limed sludge 2**
AgA	50.0 (a)***	52.8 (a)	51.7 (a)
A gB	11.3 (a)	14.7 (b)	16.4 (c)
AgE	19.7 (a)	22.1 (b)	21.5 (b)

*   Total applied 100 t DM/ha (Four applications 30, 10, 30, 30)

**   Total applied 200 t DM/ha (Four applications 60, 20, 60, 60).

***   Means on the same line followed by the same letter are not significantly different at the 5 % level.

**** Water stable aggregates after ethanol (Ag A)-benzene (Ag B) soil pretreatments and without pretreatment (Ag E).

TABLE 6 : RESIDUAL EFFECTS OF LIMED SLUDGES ON SOIL STRUCTURE

STABILITY MEASURED FOUR YEARS AFTER THE LAST APPLICATION

(1979)

Water stable aggregates ****	check soil	limed sludge 1*	limed sludge 2**
		%	
Ag A	50.0 (a)***	53.0 (a)	56.0 (a)
Ag B	11.3 (a)	13.4 (ab)	13.9 (b)
Ag E	19.7 (a)	19.4 (a)	22.2 (b)

*   Total applied : 70 t DM/ha (three applications 30, 10, 30).

**   Total applied :140 t DM/ha (three applications 60, 20, 60).

***   Means on the same line followed by the same letter are not significantly different at the 5 % level.

**** Water stable aggregates after ethanol (Ag A)-benzene (Ag B) soil pretreatments and without pretreatment (Ag E).

In order to assess the efficiency of lower applications of these products, a 10 t DM/ha rate was studied in this openfield. Results of stable aggregates measured after a 6 months period following the application are presented in table 7. The water stable aggregates showed significant increases (Ag B and Ag E) assuming that an improvement of the soil structure could be expected even for low rates of limed sludge applications.

The soil structure stability can be also reflected by measuring its permeability according to Henin and al. 1958. Results given in table 8 were obtained in 1976, then 2,5 years after the beginning of the experiment. They show that limed sludges induced an increase of the soil hydraulic conductivity.

Similar results have been obtained with unlimed sludges (12) but this type of sludge would be less effective because the modification of the soil permeability is specially related to the calcium content which contribute to flocculate soil particles.

6. CONCLUSION

The lime added to sludges mostly at high rates influences greatly the behavior of its organic matter in the soil; it appears less degradable especially in the time following the application. The soil physical properties such as structure stability beneficiate of the liming effect and of the specific type of organic matter. However, these organic matter properties are rather effective with high application doses. This fact is of particular interest for the greening of destroyed soils where a high conservation of the organic matter is expected.

TABLE 7 : INFLUENCE OF A 10 t DM/HA APPLICATION ON THE SOIL

STRUCTURE STABILITY

Water stable* aggregates	check soil	10 t DM/ha limed sludge application
	%	
Ag A	50.0 (a)**	50.5 (a)
Ag B	11.3 (a)	13.7 (b)
Ag E	19.7 (a)	22.7 (b)

*   see table 5

**   means on the same line followed by the same letter are not significantly different at the 5 % level.

TABLE 8 : INFLUENCE OF LIMED SLUDGES ON THE SOIL HYDRAULIC

CONDUCTIVITY (1976)

	check soil	limed sludge 1	limed sludge 2
	cm/h		
Permeability coefficient K	10.3 (a)*	21.7 (b)	19.7 (b)

* Means followed by the same letter are not significantly different at the 5 % level.

7. REFERENCES

(1) COLIN F. 1980 - Connaissances actuelles en matière d'utili-
     sation agricole des boues résiduaires urbaines.

     Rapport Ministère de l'Environnement et du Cadre de Vie
     n° 78 02 061.177 p.

(2) CHAUSSOD R., GERMON J.C., CATROUX G. 1978 - Détermination
     de la valeur fertilisante des boues résiduaires. Aptitude
     à libérer l'azote. Rapport Ministère de l'Environnement et
     du Cadre de Vie. 57 p. + annexes.

(3) FARDEAU J.C., GUIRAUD G., HETIER J.M. 1978 - Répercussions
     de l'épandage de boues résiduaires sur la mobilité de Cd,
     Cr, Hg, Zn et P dans les sols. 1er Symposium Sol et Déchets
     Solides. Orléans.

(4) MOREL J.L., 1978 "Effects of sludge processing on soil
     phosphorus mobility measured with the isotopic dilution kine-
     tics method. Proceedings of the EEC seminar 'Phosphorus in
     sewage sludge and animal waste". Groningen Netherlands,
     June 12 and 13. HUCKER TWG and CATROUX G. Editors.

(5) LECLERC H. 1971. Les microogranismes pathogènes des eaux
     résiduaires : évolution au cours des traitements d'épura-
     tion.

     T.S.M. 66 11 389.400.

(6) GUCKERT A., MOREL J.L. 1979. Bilan de cinq années d'utilisa-
     tion de boues résiduaires urbaines sur plantes de grande
     culture dans les conditions agroclimatiques lorraines.
     Symposium européen Cadarache 13 - 15 février.

(7) SOON Y.K., BATES T.E., BEAUCHAMP E.G., MOYER J.R. 1978.
     Land application of chemically treated sewage sludge : I
     Effects on crop yield and nitrogen availability. J. Environ.
     Qual. 7, 2 ,264-269.

(8) MOREL J.L., SEDOGO M., JACQUIN F. 1977. Biodégradation et
    humification   IV Influence de la technologie d'épuration
    sur l'évolution d'une boue résiduaire urbaine incorporée
    à un sol neutre. Bull. A.F.E.S. 3 157-166.

(9) MOREL J.L., GUCKERT A., SEDOGO M. 1978. Effets de l'épanda-
    ge de boues résiduaires urbaines sur l'état physique du
    sol. XIème Congrès de l'AISS Edmonton Canada : 19, 27 juin.

    Paru dans Bull. ENSAIA XX I II 13-19.

(10) MOREL J.L. 1977. Contribution à l'étude de l'évolution des
     boues résiduaires dans le sol. Thèse Doct. Ing. Univ. Nancy
     I 117 p.

(11) LE TACON F., CLEMENT A., GARBAYE J., MOREL J.L. 1978.
     Valorisation des boues résiduaires de stations d'épuration
     urbaines et sylviculture. Conséquences sur l'environnement.
     Rapport Ministère de l'Environnement et du Cadre de Vie.

(12) EPSTEIN E., TAYLOR J.M., CHANEY R.L. 1976. Effects of sewage
     sludge and sludge compost applied to soil on some soil
     physical and chemical properties J. Environ. Qual. Vol. 5
     4 422 426.

(13) GUIDI G. 1980 Relationships between organic matter of sewage
     sludge and physical chemical properties of soils.
     2nd European symposium on characterization, treatment and
     use of sewage sludge. 20-23 october Vienna (Austria).

(14) GUCKERT A. 1973. Contribution à l'étude de   polysaccharides
     dans les sols et leur rôle dans les mécanismes d'aggréga-
     tion.

     Thèse d'Etat Univ. Nancy I 124 p.

(15) HOHLA G.N., JONES R.L., HINESLY T.D. 1978. The effect of
     anaerobically digested sludge on organic fractions of blount
     sit loam. J. Environ. Qual. 7 4 559.

DISCUSSION

Dr GUIDI :   In your opinion which is the most important class of organic compounds, i.c. polysaccharides which are hydrophilic or oil and grease which are hydrophobic in the stabilization of soil aggregates ?

Dr MOREL :   In the soil, polysaccharides on the one hand and oils and greases on the other hand present different roles. Polysaccharides act as the initiators of the aggregation processes while oils and greases could produce an indirect protection against water degradation of the newly formed aggregates by reduction of the wettability. The stabilization of aggregates is highly dependant of the resistance of these organic compounds to biodegradation.

Dr GUIDI :   Haven't you any figures on the evolution of polysaccharides during your incubation experiments ?

Dr. MOREL :   No, I haven't.

Dr. DE HAAN : A comment. Dr MOREL mentionned rather high oil contents in sewage sludges. As far as I know it is rather difficult to determine oil in sewage sludges. I did some research work for a Dutch Oil Company about the effect of oil application to soil in crop growth and drainage watter quality. In return, chemists of the Oil Company determined for me in a number of sewage sludges the oil content, because they were specialized in determining oil. They found up to 10 % and more in sewage sludges samples of which I knew they could contain maximum 2 % of oil, because there was no negative effect of these sludges. Prof. KICK (Bonn) found out with more than 2 % of oil added to the sludges there were negative effects on yield. Therefore I think oil is determined in sludges which contain no oil. Some compounds of OM in sewage sludges may have the same characteristics typical for oil products.

Dr WILLIAMS : In sludges, it is not so much the oil that is present but probably fats and greases which might be the responsible binding agents for soil aggregates.

Dr BERGLUND : The oil concentration in sludge very much will depend upon the oil management habit in the country or region.

Dr DANNEBERG : Was the pH value of the soil changes due to the lime addition ?

Dr MOREL: Yes, it was, from 7,4 to the check soil to 7,7 in the incubated soil + limed sludge mixture.

Dr DANNEBERG : We do no longer use NaOH for extraction of soil organic matter since we found a breakdown of large molecules due to NaOH action.

Dr MOREL : You're right. The NaOH extraction is not the best technic for extracting the soil organic matter without any transformations of humic compounds. We used that method in order to compare our results to previous data obtained on natural soils and especially on calcareous soils studied by this way.

Dr GUIDI : A Comment.
The are many methods to extract soil organic matter and I think all of them produce artefacts and comparisons between results of different methods of extraction are not correct. However, I think that each of extraction methods can be used to appreciate variations of individual fractions of soil organic matter.

Dr FURRER :   How much calcium (Ca) was in the limed sludges ? And which was the pH ?

Dr MOREL :   There was about 15 % of Ca in these sludges. The pH was higher than 12 in the fresh material. After drying it fell down to 9.

Dr CATROUX   :   You observed an important O. Matter decrease in your control plot. Have you any explanation for that ?

Dr MOREL :   The decrease was about 0.2 % of organic carbon. This is high and it was probably due to the fact that in these plots no organic, no mineral fertilization were made. However, artefacts related to the depth of the soil sampling or the soil ploughing could have modified the carbon concentration in the samples.

Dr SÜSS :   A comment. With limed sludge the effect on soil is a complex one. Lime alone has an effect on physical properties of the soil. The effect of organic matter and lime cannot be derided.

SOME EFFECTS OF SEWAGE SLUDGE ON SOIL PHYSICAL

CONDITIONS AND PLANT GROWTH

J.E. HALL AND E.G. COKER

1. INTRODUCTION AND REVIEW OF LITERATURE

Soils in the process of reclamation or those in continuous cultivation are often very deficient in organic matter. The cultivation process speeds up the decay of soil organic matter and
the organic matter returned from crop residues is less than that
returned to the soil from a full vegetation cover. The use of
heavy machinery often causes soil damage due to excessive com-
paction.

Soils lacking organic matter are more likely to suffer problems
arising from lack of structure (1). This is because soil organic
matter binds together soil particles into aggregates between
which there are large (non-capillary) pore space through which
air can penetrate into the soil and through which surplus water
can drain away. The smaller pores between individual soil
particules are more or less filled with water when the soil is
moist and air does not move freely in them. Cultivations break
soil up into lumps or aggregates, but where organic matter or
clay is lacking, the soil aggregates are not very stable in the
presence of rain or percolating water (i.e. the aggregates lack
water stability) and easily disintegrate into their constituent
particules. When this happens the larger soil pores are lost,
the volume of soil air is reduced, the rate of movement of rain
or irrigation water becomes slower and in extreme conditions
stops. The soil becomes more closely packed and the bulk
density is increased. Thus the aggregate stabilities of sandy
and silty soils with a low clay percentage are particularly
sensitive to lack of soil organic matter. Such soils frequently

suffer structural problems when the organic matter content is below 3 %. The organic content of most mineral soils rarely exceeds 6 %.

Fertile soils should contain not less than 10-15 per cent by volume of non-capillary pores through which air and surplus water can move freely. Such soils will have lower bulk densities than over-compact soils. Thus bulk density in soil is related to other measurements of soil structure and forms a useful index for the assessment of structural change in soil.

The addition of organic manures to sands, silts and silt loams in particular have improved soil properties so that crop yields have been increased even when optimum amounts of conventional fertiliser have been applied (2).

Organic matter from cattle manure increases the total and air-filled pore space and reduces bulk density, particulary in the topsoil (0-15 cm); it also increases the water stability of soil aggregates (3). The depth and density of crop root growth has been increased by farmyard manure (FYM)(4). Increases in organic matter from FYM also increase the available soil water holding capacity; almost all of this increase is readily available to crops (5).

Johnston (6) stated that the long-term physical effects of organic matter contained in sewage sludge were no different from those of FYM. Bunting (7) found that digested sludge solids were more effective in increasing the water-holding capacity of soil than FYM but less effective in increasing crop production. These observations can be reconciled if it is remembered that FYM decays rapidly after being mixed into soil and releases most of the plant nutrients during the first two years after application leaving a stable residue or humus which is analagous to the stable organic residue of digested sludge. It is there-fore to be expected that the addition of sewage sludge to soil modifies soil properties in similar ways to FYM.

For example, increases in the content of organic matter in top-soil following rotary cultivation of 50-60 tonnes sludge solids/ha were found to be about 1.6% (8,9).

At application rates of 50-60 tds/ha sludge the amount of water at soil tensions between 0.33 and 15 bars (i.e. field capacity to permanent wilting point) was increased from 12.5 to 18.5 per cent (8,9,10), and was readily available to plants. Nine years application of sewage sludge raised the available water capacity of a sandy loam by 35 per cent while FYM raised the available water capacity by 30 per cent (11). On a fine sandy loam this is equivalent to 100-140 m^3 water/ha, (i.e. 10-14 mm rain).

The addition of sewage sludge solids increased the proportion of soil particles as stable aggregates from 16 per cent to 33 per cent of the total (12). Kladivko and Nelson give data quoted in Table 1 (9).

These results indicate that where sewage sludge has been added to soil in continuous cultivation, porosity can be increased and bulk density reduced to values characteristic of fertile virgin soil under permanent vegetation cover where soil organic matter levels are high. The reduction in bulk density occurred as a result of increase in the non-capillary pore space. Comparison of the data for November and May shows the marked effect of exposure of bare soil to winter rainfall and demonstrates that sludge-treated soil became less compacted than untreated soil.

TABLE 1 : MEAN VALUES OF SOIL PROPERTIES IN THE 0-5 cm LAYER
FOLLOWING APPLICATIONS OF 56-80 TDS/HA SLUDGE (9).
(SIX MONTHS AND 12 MONTHS FROM TIME OF SLUDGE
APPLICATION REPRESENT APPLICATIONS IN NOVEMBER AND
MAY).

Treatment	November		May	
	Untreated	Treated	Untreated	Treated
Bulk density	1.19	1.00	1.29	1.13
Mean weight diameter of soil aggregates, mm	0.84	1.82	0.46	1.29
Per cent volume of non-capillary pore space	19.2	27.8	11.4	16.8

An increase of soil porosity will lead to an increase in the
rate of infiltration of water into soil, and hence reduce water
loss and erosion by surface run-off. Where liquid sludge is applied to soil, the soil pores are temporarily sealed and the
rate of water infiltration is reduced for a few days. As soon
as the sludge layer begins to dry it will crack and water enters
freely between the fragments. The sludge fragments for a while
protect the underlying soil from structural breakdown and
blockage of the soil pores. Thus the longer term effect is to
increase the possible water infiltration rate into soil. Sludge
solids are pulled into the soil by earthworms as they begin to
disintegrate, and earthworm burrows are of prime importance for
the rapid infiltration of surface-applied water (9). Infiltration rates increased on a sandy loam by about 50 per cent where
56 tds/ha sludge were applied compared with none but there was
no effect on a silty loam. These effects lasted for at least
two years (13).

In the experiments reported below soil physical properties have
been measured on soils which received sewage sludge up to four
years previously and on which grass and other crops have been
grown.

2. EXPERIMENTAL DETAILS

2.1. The Experiments

The results reported here were taken from three existing series
of experiments which were investigating the manurial value
(Experiments A and B) and the uptake of potentially toxic
elements by crops (Experiment C), of several kinds of sewage
sludge.

2.1.1. Experiment A compared the effects of lime-copperas and
polyelectrolyte treated undigested sewage sludge on
grass grown on three soil types; silt loam (organic matter OM
content 2.1 %), sandy loam (OM 2.9 %) and clay loam (OM 3.8 %).
Soil and root samples were taken three years after sludge incorporation.

2.1.2. Experiment B compared grass yields produced by air-dried
digested sludge and undigested sludge cake when they
were rotary cultivated into the boulder clay covering a re-
claimed refuse tip. Soil bulk density measurements were made
four years after sludge incorporation.

2.1.3. In Experiment C, lagoon-matured digested sewage sludge
was applied to a calcareous loam. Liquid sludge was applied at rates of up to 152 t/ha dry solids; application rates
were extended up to 500 t/ha by using air-dried digested sludge
from the same source. Soil density and water holding capacity
were measured two years after the application of liquid sludge,
and one year after the solid sludge was applied.

2.1.4. Sludge solids applications.
These are given in Table 2.

TABLE 2 : RATES OF APPLICATION OF SLUDGE IN EXPERIMENTS A, B and C (tds/ha)

| Experiment A | | Experiment B | | Experiment C | |
Lime copperas undigested	Polyele- trolyte undigested	Air dried digested cake	Undigested cake	Liquid digested	Air dried digested
0	0	0	0	0	
27	27	35	20	19	
53	53	70	160	38	
80	80	140		76	
107	107	280		152	
					250
					500

2.2. Methods

Soil water holding capacity and bulk densities were determined by standard MAFF methods (14). Root samples were taken with a needle board at depths of 0-150 mm and 151-300 mm (15). Carbon and organic matter contents were determined by a standard MAFF method (16) and nitrogen was determined by the published WRC method (17).

3. RESULTS

3.1. Effect on soil Density

3.1.1. Experiment A

Significant reductions in soil bulk densities were found in the silt loam, sandy loam and clay loam soils. There were no signi-

ficant differences between the effects of the lime-copperas and
the polyelectrolyte treated solid undigested sludges (Table 3).

TABLE 3 : EXPERIMENT A.   SOIL BULK DENSITY g/cm^3

L-Lime copperas sludge; P-Polyelectrolyte treated
Figures are means of 16 samples

Sludge applied tds/ha	Silt loam		Soil type Sandy loam		Clay loam	
	L	P	L	P	L	P
0	1.18	1.48	1.43	1.43	1.64	1.64
27	1.43	1.35	1.24	1.27	-	-
53	1.41	1.31	1.28	1.22	-	-
80	1.13	1.13	1.24	1.29	-	-
107	1.10	1.04	1.18	1.28	1.50	1.55
133	1.07	0.89	1.16	1.22	1.54	1.53
LSD	0.12	0.11	0.10	0.11	0.06	NS

The silt loam soil was analysed for nitrogen, carbon and organic
matter; each was increased as the level of sludge solids additions became larger (Table 4). The organic matter content of
the unsludged soil increased by 0.14 % following the growth of
grass for three years. However the differences between treatments have probably been reduced as a result of the decay of
about 80 % of the applied organic matter over this period.

At four out of the five levels of application, the soil organic
matter content has been increased by 0.008-0.010 per cent per
tonne of sludge solids applied; the increase at the fifth rate
(80 t/ha) was 0.013 per cent per tonne. The theoretical addition
was 0.04 per cent per tonne.

TABLE 4 : ANALYSIS OF SILT LOAM SOIL THREE YEARS AFTER SLUDGE
APPLICATION

Sludge applied tds/ha	% OM	% C	% N	C:N
0	2.24	1.30	0.22	5.91
27	2.49	1.44	0.25	5.76
53	2.76	1.60	0.28	5.71
80	3.04	1.76	0.32	5.50
107	3.11	1.80	0.30	6.00
133	3.34	1.94	0.36	5.38

3.1.2. Experiment B

Both the air dried digested sludge and the undigested sludge
cake significantly reduced the bulk density of the compacted
boulder clay (Table 5).

TABLE 5 : EXPERIMENT B. SOIL BULK DENSITY g/cm^3
FIGURES ARE MEANS OF 16 SAMPLES

Air dried digested sludge		Undigested sludge cake	
Sludge applied tds/ha	Density	Sludge applied tds/ha	Density
0	1.78	0	1.78
35	1.79	20	1.76
70	1.75	160	1.51
140	1.70		
280	1.46		
LSD	0.11	LSD	0.10

3.1.3. Experiment C

The bulk density of the calcareous loam was reduced by both
liquid and air-dried digested sludge solids (Table 6).

TABLE 6 : EXPERIMENT C. SOIL DENSITY g/cm^3
FIGURES ARE MEANS OF 8 SAMPLES

Sludge applied tds/ha	Lagoon digested sludge
0	1.13
19	1.10
38	1.11
76	1.09
152	0.99
250	0.87
500	0.72
LSD	0.07

3.2. Effects on root growth

In Experiment A grass root samples were taken from two depths of
the silt loam soil and they showed increased root growth at the
higher sludge rates; there were fewer roots between 151-300 mm
(Table 7). The roots were closely associated with sludge
particles.

TABLE 7 : GRASS ROOT DRY WEIGHTS mg/cm^3
SOIL

	Lime-copperas undigested sludge tds/ha			Polyelectrolyte undigested sludge tds/ha		
Depth mm	0	107	133	0	107	133
0-150	1.43	2.38	3.13	1.43	3.58	2.30
151-300	0.13	0.04	0.04	0.13	0.07	0.05
0-300	1.56	2.42	3.17	1.56	3.65	2.35

3.3. Effects on soil water holding capacity

In Experiment C, the field capacity of soil from treated plots was measured after settlement following cultivation (Table 8). The results show that the water content at field capacity was increased by sludge additions.

TABLE 8 : EXPERIMENT C. SOIL WATER HOLDING CAPACITY
         FIGURES ARE MEANS OF 8 SAMPLES

Sludge applied tds/ha	% moisture	Water holdings capacity ml/cm^3 soil
O	30.1	0.340
19	32.2	0.354
38	32.4	0.360
76	35.2	0.384
152	39.4	0.390
250	43.1	0.375
500	47.3	0.340
LSD	1.8	

4. DISCUSSION

Continuous arable cultivation returns little organic matter to the soil as crop residues or farmyard manure and can result in loss of soil structure and increased soil density making culti- vations more difficult and the soil more prone to physical damage (1). There may be deleterious effects on drainage, water holding capacity and aeration, all of which can contribute to declining crop yields. It is well documented that the addition of organic matter will ameliorate these conditions and improve yields. Farmyard manure contains about 74 % of organic matter on a dry solids basis compared with about 55-70 % in sewage sludge depending on whether it has been digested (19).

Significant reductions in soil bulk density have been achieved with an application rate of 27 tonnes/ha sludge solids of undigested sludge on the three soil types investigated in Experiment A (Fig. 1). Application rates of this order are commonly used for FYM. At this rate, the greatest effect was found in the sandy loam, but the largest overall reduction in density was in the silt loam. The clay loam showed the least reduction in density but its friability was greatly improved at the highest application rate; when treated plots were ploughed, the soil crumbled and was not smeared by the mouldboard.

The organic matter in the silt loam soil was below the deficiency level of 3 % but this was progressively alleviated with each sludge increment (Table 5). To raise the organic matter concentration of this soil above 3 % required 80 tds/ha of sludge. The increased root density reflects the improved fertility of the soil (Table 7). The C:N ratio was much lower than the commonly found ratio of 10:1 and decreased with sludge addition (Table 4).

When boulder clay is used in the reclamation of municipal refuse tips, it commonly becomes very compacted; it has a very low organic matter content. In Experiment B it was found that much larger quantities of sludge were required to significantly reduce soil density than in Experiment A (Fig. 2). The undigested sludge cake was twice as effective as the air-dried digested sludge.

The bulk density of the calcareous loam was reduced by the addition of sludge solids but it required larger quantities to achieve a similar reduction in soil density to that in Experiment A. This was due to the already open structure of the unamended soil. The water content of the soil at field capacity significantly increased with each sludge addition (Fig. 3). Lettuce yields for 1979 and 1980 also increased as the amount of sludge solids added became larger, even though all plots received sufficient N, P and K for full growth (Fig. 4). This

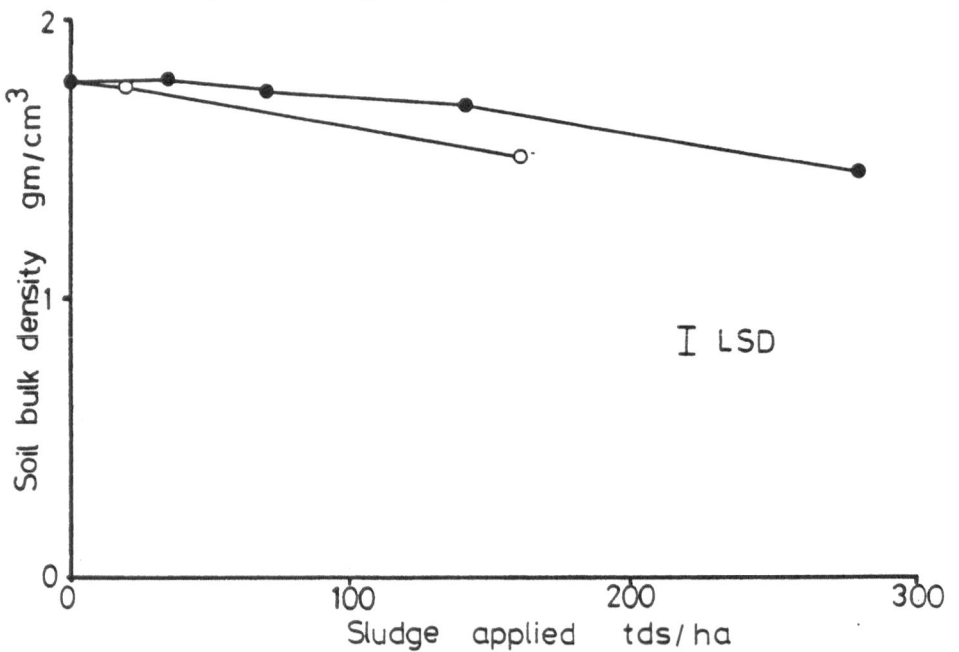

Figure 1 : The effect of solid undigested slude on soil bulk density. Experiment A. Points are means of both sludge types. Δ clay loam, O silt loam, ● sandy loam

Figure 2 : The effect of air dried digested sludge (●) and undigested sludge cake (O) on the bulk density of compacted clay. Experiment B.

Figure 3 : The effect of digested sludge solids on the bulk
density and moisture content of calcareous loam.
Experiment C. ● % moisture, O bulk density

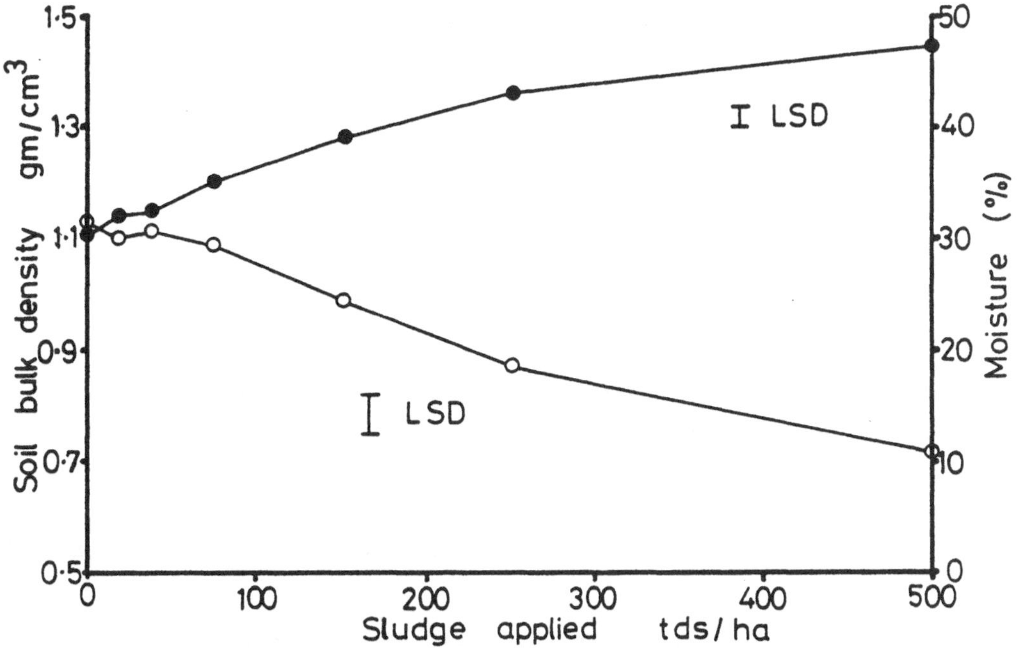

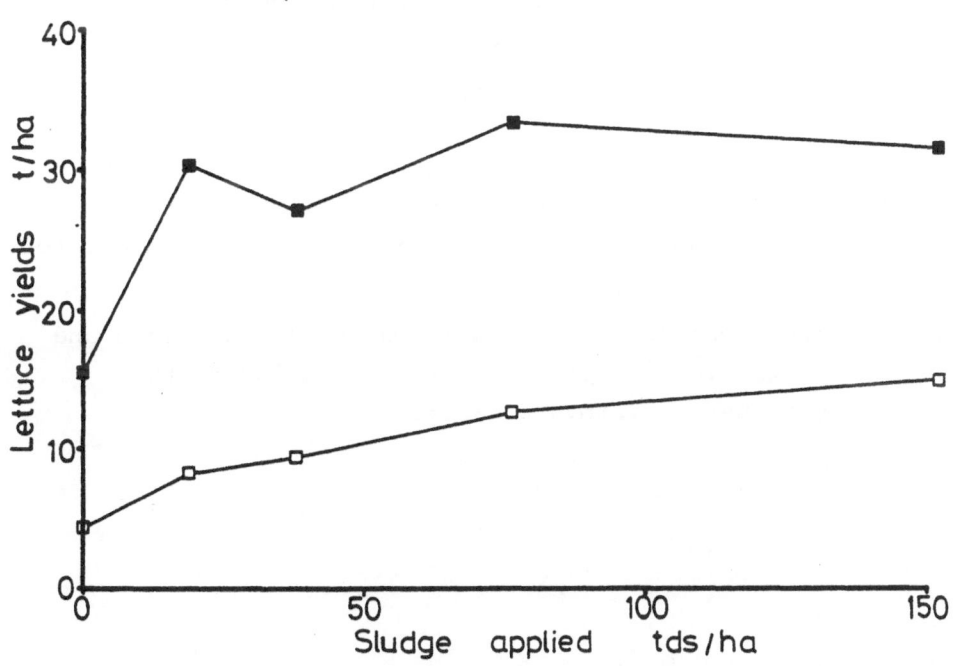

Figure 4 : Lettuce yield t/ha fresh weight. Experiment C.
□ 1979, ■ 1980.

was presumably due to the increased water holding capacity of the sludged soil. Rainfall for the June–August period 1980 was 191 mm, double that for the similar period in 1979. The effect of sludge on crop yield was more marked in the drier year suggesting that improvements in soil water holding capacity following application of sludge can help to compensate for droughty conditions.

5. CONCLUSIONS

1. Bulk density is a useful indicator of soil structure. Significant reductions in density have been achieved on three arable soils following the application of undigested sewage sludge at rates in excess of 27 tds/ha. Polyelectrolyte treated sludge was equally effective as lime-copperas sludge.

2. The bulk density of a boulder clay used on a land reclamation site was reduced but required higher rates of sludge application than for arable soils. Undigested sludge cake was twice as effective as air dried digested sludge.

3. Digested sludge solids decreased the bulk density of calcareous loam and increased its water holding capacity. In a year of low summer rainfall, lettuce yields were doubled by the addition of 19 tds/ha sludge even when sufficient fertiliser N, P and K had been applied to all treatments for full growth.

4. Sewage sludge solids improved soil physical conditions and grass root growth at application rates similar to the agricultural use of farmyard manure.

6. SUMMARY

There is evidence to suggest that because of their beneficial effects on soil physical properties, organic manures can improve crop yields more than inorganic fertilisers of equivalent plant nutrient content. The experiments reported below have investigated the possibility that organic matter in sewage sludge is also beneficial. Various physical properties of soils have been measured following applications of sludge made up to four years previously, and the results have been compared with crop yields. Soil bulk density was decreased by undigested sludge applied at 27 tonne dry solids (tds)/ha and soil water holding capacity was significantly increased by the addition of $\geqslant$ 19 tds/ha of digested sludge. On a calcareous soil applications of sludge increased lettuce yields by more than 100 per cent despite generous applications of inorganic fertiliser to control plots. It is thought that improved soil water holding capacity on the sludged plots was the principal cause of these substantially improved yields. Beneficial effects attributable to sludge organic matter were of the same order as those reported for conventional organic fertilisers and became evident following applications of about 20-30 tds/ha (equivalent to about 14-21 t sludge organic matter per ha), rates comparable to those used for farmyard manure.

7. ACKNOWLEDGMENT

As part of its comprehensive research programme on sewage sludge disposal, Experiment C is funded by the Department of Environment. The Water Research Centre gratefully acknowledges this funding and the permission of the Department to publish some of the results in this paper.

8. REFERENCES

(1)   STRUTT, N. Modern Farming and the Soil. Report of the Agricultural Advisory Council on Soil Structure and Soil Fertility. MAFF, Her Majesty's Stationery Office, London, 1970.

(2)   DE HANN, S. Humus, its formation, its relation with the mineral part of the soil, and its significance for soil productivity. Soil Organic Matter Studies, 1, 21-30. (Proceedings of a Symposium IAEA-FAO, Braunschweig).

(3)   MAZURAK, A.P., CHESMIN, L., and THIJEEL, A.A. Effects of beef cattle manure on the water stability of soil aggregates. Soil Science Society of America. Journal, 41, 1977, 613-615.

(4)   HAMBLIN, A.P., and DAVIES, E. Influence of organic matter on the physical properties of some East Anglian soils of high organic matter content. Journal of Soil Science, 28, 1977, 11-22.

(5)   SALTER, P.J., and WILLIAMS, J.B. The effect of farmyard manure on the moisture characteristics of a sandy loam soil. Journal of Soil Science, 14, (1), 1963, 73-81.

(6)   JOHNSTON, A.E. Research seminar on the manurial value of sewage sludge, November 1976, p. 3-11. Directorate Water Engineering Research and Development Division Technical Note No. 9. Department of the Environment, 1977.

(7)   BUNTIN, A.H. Experiments on organic manures, 1942-9. Journal of Agricultural Science Cambridge, 60, 1963, 121-140.

(8)   EPSTEIN, E., TAYLOR, J.M., and CHANEY, R.L. Effects of sewage sludge and sludge composts applied to soil on some physical and chemical properties. Journal of Environmental Quality, 5, (4) 1976, 422-6.

(9) KLADIVKO, E.J., and NELSON, D.W. Changes in soil properties from application of anaerobic sludge. Journal of the Water Pollution Control Federation, 51 (2), 1979, 325-332.

(10) GUPTA, S.C., DOWDY, R.H., and LARSON, W.E. Hydraulic and thermal properties of a sandy soil as influenced by incorporation of sewage sludge. Soil Science Society of America Journal, 41, (3) 1977, 601-605.

(11) SALTER, P.J., WILLIAMS, J.B., and HARRISON, D.J. Effects of bulky organic manures on the available water capacity of a fine sandy loam. Experimental Horticulture, 13, 1965, 69-75.

(12) EPSTEIN, E. Effect of sewerage sludge on some soil physical properties. Journal of Environmental Quality, 4, 139-142, 1975.

(13) KELLING, K.A., PETERSON, A.E., and WALSH, L.M. Effect of wastewater sludge on soil moisture relationships and surface run-off. Journal of Water Pollution Control Federation, 49, (7), 1977, 1698-1703.

(14) ARCHER, J.R., and MARKS, M.J. Techniques for measuring soil physical parameters. ADAS advisory paper No. 18, MAFF, London.

(15) GOEDEWAGEN, M.A.J., (Die methoden, die aan het Landbouwproef-station.) The techniques of root investigation in cultivated and grassland used at the Soil Science Institute, Groeningen. Landbouwproefstation-Bodemkundig Instituut TNO Groningen, Netherlands, 1948.

(16) MINISTRY OF AGRICULTURE, FISHERIES AND FOOD. The analysis of agricultural materials. Technical Bulletin 27, HMSO, London, 1973.

(17) PETTS, K.W., and BELCHER, M. The simultaneous determination of nitrogen and phosphorus in plant material. WRC, LR 1033, 1979.

(18) MINISTRY OF AGRICULTURE, FISHERIES AND FOOD. Making the most of farmyard manure. ADAS Leaflet 435, 1979.

(19) BECKETT, P.H.T., DAVIS, R.D., MILWARD, A.R., and BRINDLEY, P., A comparison of the effect of different sewage sludges on young barley. Plant and Soil, 48, 1977, 129-141.

DISCUSSION

Dr CATROUX :   Did you take in account a possible phosphorus effect in your trials ? Why can you distinguish between P and organic matter effect ?

Dr HALL :   All plots received sufficient P fertilizer to remove any effect of sludge P.

Dr DIEZ :   In contrary to our results you have found, that the decay of organic matter increases with the amount of sludge applied. Can you explain that ?

Dr HALL :   The apparent differences in the decay rates of low and high rates of applied sludge are from 77 % to 82 %. However these are not significantly different.

Dr WILLIAMS :   A comment.
Soluble nitrogen in liquid digested sludge appeared during the winter period would have been lost in the following spring, giving a low N value for the succeeding crops. In the 2nd year, the effect would be better as a result of N release following organic matter of composition.

Dr BARIDEAU :   What were the dimensions of the plots in the field trials. Don't you think they were somewhat narrow ?

Dr HALL :   Width of plots varied between 1.3 m and 1.8 m and length between 10 m and 20 m depending on the experiment. Solid sludges were rotavated along the length of the plots so lateral movement were negligible. Liquid sludges were applied to plots which had had an impermeable barrier placed 30 cm into the ground to prevent movement.

GENERAL COMMENTS ON THE ORGANIC

VALUE OF SLUDGE

S. DE HAAN

At the end of the first session, after the paper presented by
Dr MOREL, the chairman, Dr CATROUX remarked that farmers were
not interested in effects on soils, but on yields.

In my opinion, the most interesting effects of organic manures,
among them sewage sludge, is not the direct (or short-term)
fertilizer (NPK) effects, but the so-called organic matter effect. That is an increase in soil productivity which cannot
be effectuated by mineral fertilizers only. This organic matter
effect can only be measured in field experiments with increasing
application rates of mineral nitrogen, including the optimal
application rate, which cannot be determined beforehand, because
of possibilities of leaching and volatilization of nitrogen.

In long-term field experiments in the Netherlands there was a
clear positive organic matter effect, an other there was none.
The reason for these differences in reaction to organic manures
are not (yet) known. The number of long-term field experiments
to this question is too small.

Some years ago six experimental fields with sewage sludge were
laid out on which also this organic matter effect can be studied.
It is possible that on these experimental fields "the long run
negative organic matter" effects will appear because of high
levels of heavy metals in sewage sludges. On some sensitive
(= acid sandy) soils with long-term application to town refuse
compost such a suspective effect was observed with crops sensitive to excess of heavy metals.

In this context DE HAAN showed also some results of his experiments about humus formation from different kinds of organic matter and on different soils. From these experiments up to now could be concluded that the content of lignin -like substances in the organic manures determined the amount of humus formed. Differences between different soil types were not clear but all (36) soils were kept under the same conditions in this experiment. In practice there may be more wet conditions in heavy soils and dry conditions in light sandy soils. DE HAAN showed the effect of different sources of organic matter on soil structure. All kinds of organic matter have favorable (short-term) effects on soil structure, irrespective of the amount of humus formed by them. The latter has a more long-term effect on soil structure.

DISCUSSION

Dr HALL :   You say that decomposition of organic matter
             of sewage sludge is the same as FYM. Is it
             not that FYM decays rapidly $\simeq$ 50 % in the year
             of application and sludge decays very much
             slower because it is already in a relatively
             stable form ?

Dr DE HAAN : If I have said that it is not quite correct.
             I had no time to collect my data on decomposi-
             tion of organic matter from sewage sludges, but
             I know it is different from sludge to sludge.
             For most sludges, I think decomposition is in
             the same order of magnitude as for farm yard
             manure, which has passed also several decom-
             position processes before it is applied to
             land. But, I know in some sludges organic matter
             is practically undecomposable whereas in other
             it may be more decomposable than in FYM. The
             reason for this difference is not yet clear.
             I think it is the degree of stabilization of
             the sludges which makes the difference, but it
             is difficult to define the degree of stability
             of sludge. Under the same conditions ash con-
             tent might be a measure for stability, but ash
             content may be increased by other reasons than
             mineralization. For example, by addition of
             lime or ferric and/or aluminium salts to the
             sludges. And maybe also heavy metals affect
             the decomposability of sludge.

INFLUENCE OF SEWAGE SLUDGE APPLICATION ON PHYSICAL

PROPERTIES OF SOILS AND ITS CONTRIBUTION TO THE

HUMUS BALANCE

FURRER, O.J. and STAUFFER, W.

1. INTRODUCTION

Almost half of the solids of sewage sludge consists of organic compounds. So sewage sludge is expected to be an important source of organic matter for humus supply and acts as soil conditioner. In addition, sewage sludge contains a large amount of calcium, another soil conditioner.

On the other hand, sewage sludge is usually rich in phosphorus. If a reasonable dose of P is applied per ha the contribution of sludge to the humus of the soil is small. In Switzerland, the mean yearly application rate per ha is limited to 2.5 t sludge dry matter by the sewage sludge ordinance.

The main objectives of this paper are as follows :

- to discuss the distribution of organic matter, Ca and P in sewage sludge samples,

- to present results of field and lysimeter experiments on the influence of sewage sludge application on aggregate stability and bulk density of soils,

- to estimate the contribution of sewage sludge in the total humus balance of Swiss soils.

2. ORGANIC MATTER, Ca AND P CONTENT IN SEWAGE SLUDGE

Organic matter (OM), Calcium (Ca) and Phosphorus (P) are the most important components of sludge influencing the physical properties of the soil. In Fig. 1 to 3, the results of OM, Ca and P determinations in 1511 sewage sludge samples of all Swiss waste water treatment plants serving more than 10,000 inhabitants from the period 1978 to 1980 are presented.

Fig. 1 shows a very large variation of OM-contents in a rather normal distribution. The trimmed mean is 41 % OM in the dry matter (DM) and is rather low as compared to the fresh sludge samples. The reduction is due to the very effective anaerobic digestion in most of the Swiss waste water treatment plants.

Figure 1 : Distribution of organic matter contents in 1511 sewage sludge samples collected during 1978-80 (7).

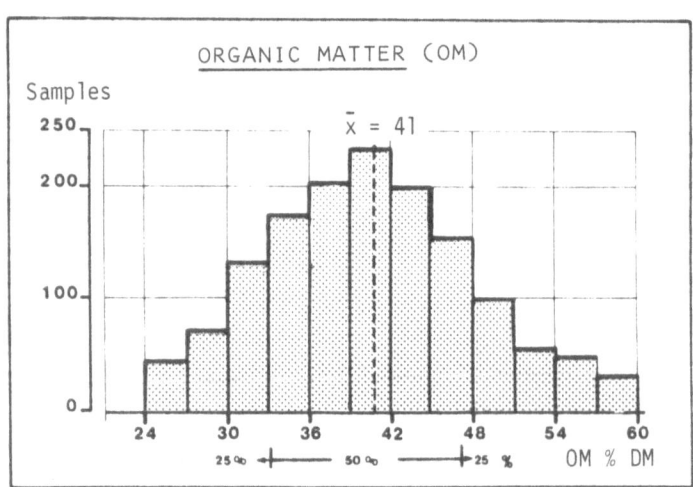

In Fig. 2, the values for Ca content are presented. The distribution of Ca is similar to that of organic matter (Fig. 1). The trimmed mean amounts to 73 g Ca per kg DM, equivalent to 18 % of $CaCO_3$ in DM. That is almost half the organic matter content.

It can be suggested that Ca is as important as organic matter for the favorable effects of sewage sludge on soil physical properties.

Figure 2 : Distribution of calcium contents in sewage sludge samples collected during 1978-80 (7).

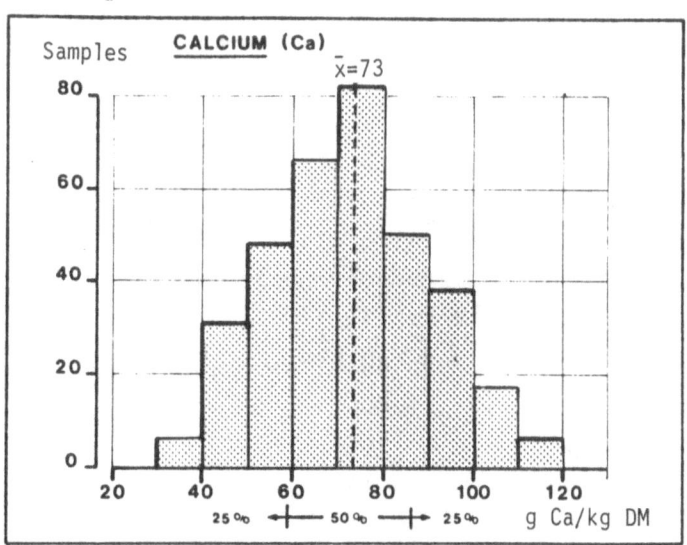

Figure 3 : Distribution of phosphorus contents in 1511 sewage sludge sample collected during 1978-80 (7).

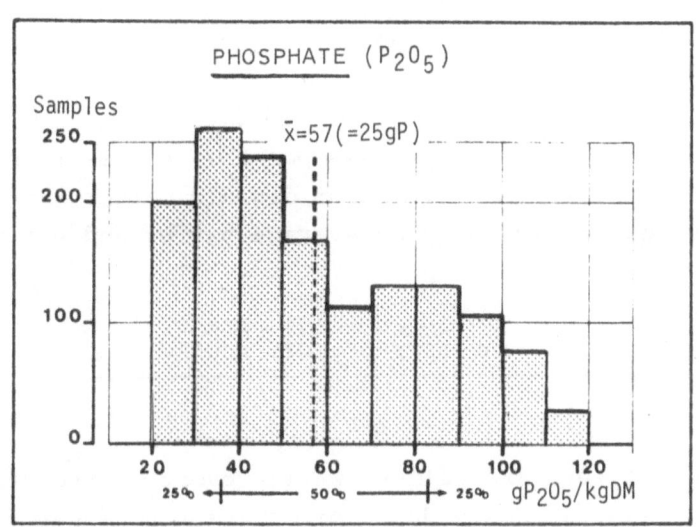

The distribution of P in sewage sludge samples is presented in
Fig. 3. The trimmed mean is 25 g P/kg DM. However, the distribution pattern is quite different from that of organic matter
and calcium. The observed difference is mainly   due to the fact
that the results presented in Fig. 3 comprises sludge samples
of waste water treatment plants with and without P-elimination
facilities. The separate averages of P-contents are 40 g and
15 g P/kg DM, respectively (6).

Sewage sludge is an important source of P, specially that obtained from waste water treatment plants having facility of P-
elimination. In this cases the ratio organic matter to P is
quite narrow (10:1). It is also clear from the data presented
below that the ratio is extremely narrow in sludge as compared
to other organic fertilizers (3) :

saw dust	10,000	:	1 (OM:P)
straw	1,000	:	1
cow slurry	100	:	1
sewage sludge	10	:	1

The fact that sewage sludge is so rich in phosphorus gives an
other aspect. If a reasonable dose of P is applied per ha, the
contribution of sludge to the humus of the soil is small, in
most cases less than 500 kg of organic matter per ha. In
Switzerland, the mean yearly application rate is limited to
2.5 t sludge DM per ha (Sewage Sludge Ordinance, in force since
Mai 1, 1981).

The recommended mean yearly doses per ha are less than 1 t
sludge DM. So, less than 400 kg of organic matter are applied
in form of sewage sludge on an average per ha and per year.

3. INFLUENCE ON THE STABILITY OF SOIL AGGREGATES
——

Organic fertilizers are well known improvers of the stability
of soil aggregates. Heavy doses of sewage sludge were applied

in field and lysimeter experiments to different types of soil
in order to assess its influence on the stability of soil
aggregates. Only in few cases such an effect was observed.

The results of a field experiment in Büren on a heavy soil (41 %
clay, 4.5 % humus, pH 7.9) are presented in Fig. 4. The trial
was started in 1975 and divided in four parts with parallel
shifted rotations. On plots receiving sewage sludge, each
year 5 t OM/ha (total 25 t OM/ha in five years) were applied.
The aggregate stability was measured 1979 after the harvest of
the last crop.

Figure 4 : Aggregate stability of a soil as influenced
           by fertilization and   by crops :
           0 = no fertilization   G = gras + clover
           M = mineral fertilizers   C = corn
           S = sewage sludge   W = wheat

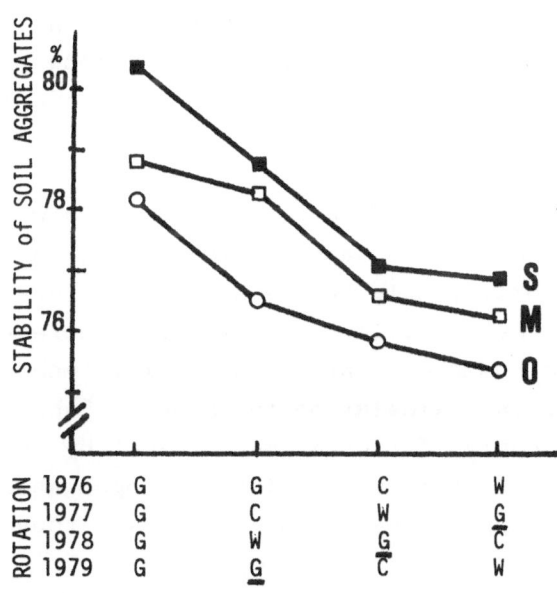

The datas in Fig. 4 show that the influence of the type of crop is of major importance. There is no significant difference between mineral and organic (sewage sludge) fertilized plots. The lower stability of soil aggregates in the control (no fertilization) is certainly due to the reduced crop growth in these plots. The most stable soil aggregates are found in plots having a permanent grass-clover-mixture. Grass has an especially favorable influence on the stability of soil aggregates. The longer the distance to the last grass crop in the rotation, the less stable the soil aggregates (Fig. 4).

The same type of field trial as presented in Fig. 4, but on a sandy loam (16 % clay) and on a loamy sand (5 % clay) did not give any effect of sewage sludge application on soil aggregate stability; however, the effect of the crops was as obvious as in the trial on the heavy soil (Fig. 4).

In Table 1 the results of an other field experiment on a heavy soil (61 % clay, 8 % humus, pH 7,2) are presented. From 1971 to 1975 two different doses of sewage sludge (120 and 480 m^3 per ha and year containing 8 % dry matter with 400 g organic matter per kg DM) were applied to permanent grassland (meadow plots) and to a rotation : wheat - corn - beets - wheat - corn (plots in rotation).

In meadow plots a better stability of soil aggregates was found than in plots in rotation. Sludge applications has a favorable effect on the soil stability only in the plots in a rotation, but not in permanent grassland. Sludge application effects a bigger increase of the humus content of the soil in meadow plots than in plots in rotation. Calculating the humus balance we made an estimation of the supply of organic matter (OM) by the crop (roots, residues) proportional to the yield of tops. About 50 % of the organic matter supplied by sewage sludge was found to be decomposed in the soil.

TABLE 1 : AGGREGATE STABILITY (%), HUMUS CONTENT (%) AND
HUMUS BALANCE IN THE TOP LAYER OF THE SOIL AS IN-
FLUENCED BY REPEATED (FIVE YEARS) SEWAGE SLUDGE
APPLICATIONS

Treatment	mineral fertilization	sewage sludge	
		120 m^3	480 m^3
Aggregate stability (%)			
meadow plots	64	64	65
plots in rotation	51	55	59
Humus content (%)			
meadow plots	8.3	8.8	10.2
plots in rotation	8.0	8.3	9.0
Humus balance (t OM/ha) for meadow plots			
a) soil content 1971	164	164	164
(0-15 cm) 1975	166	176	204
increase	2	12	40
b) supply : sewage sludge	0	19	77
plant residues	20	20	25
total (5 years)	20	39	102
c) decomposition : (b-a)	18	27	62
% of the total supply	90 %	69 %	60 %
% of OM of sludge	-	47 %	51 %

In a lysimeter experiment (2) on sandy loam (18 % clay, 4 %
humus, pH 6.2) from 1973 to 1977 yearly three different doses
of sewage sludge were applied : 4, 8 and 12 t DM per ha. The
lysimeters were planted with grass. The stability of the soil
aggregate, measured at the end of the trial, was as follows :

without fertilization	65.4 %
mineral fertilization	65.6 %
sewage sludge, 4 t DM/y	68.3 %
sewage sludge, 8 t DM/y	70.2 %
sewage sludge, 12 t DM/y	74.7 %

In this lysimeter experiment a significant, very pronounced
improvement of the stability of soil aggregates by sludge
application was observed. But it has to be mentioned, that at
the same time, with the highest sludge rate, an amount of
1365 kg P/ha was applied and an enrichement of the soil by more
than 1000 kg P/ha was attained. This quantity of phosphorus
accumulated in the soil is more than crops will need for
further 20 years (4).

In all field and lysimeter experiments mentioned in this chapter,
measurements of bulk density and of volume and size of pores
were made. In no case a significant influence of sewage sludge
application on these parameters was observed. Only in few
cases a tendency of a reduction of the bulk density induced by
sewage sludge application was observed. In very few cases only,
an augmentation of the volume of pores was found.

Due to the dark colour of the soils receiving sewage sludge
treatments higher soil temperatures at the surface of bare
soils were recorded.

In the field trials frequent attacks of weed bulb fly (Hylemia
coarctata Fall) on the sludge treated wheat crop were observed.
This harmfull insect prefers to deposit its eggs on dark colored
soils. In repeated cases it was very conspicuous that only plots
with sewage sludge application were attacked.

4. CONTRIBUTION OF SEWAGE SLUDGE TO THE HUMUS BALANCE

The agricultural land of Switzerland amounts around a million hectares. More than 70 % are used as greenland, the rest as arable land. There is an estimation (5), that this land surface is supplied with the following amounts of organic matter (OM) from different sources :

plant residues, roots	> 3000 kt OM
manure : animal excreta	2500 kt OM
straw, litter	700 kt OM
peat	50 kt OM
sewage sludge	50 kt OM
town refuse compost	20 kt OM

$(kt/Switzerland = 10^6 \ kg/10^6 \ ha = kg/ha)$

The most important amounts of organic matter originate from plant residues and manure. This is due to the preponderant portion of greenland and to the big number of animals held, partially fed with imported food. The contribution of sewage sludge to the supply of the soil with organic matter amounts to less than 1 % and is of little importance.

5. SUMMARY

Organic matter, calcium and phosphorus are the most important components of sewage sludge influencing the physical properties of the soil. The trimmed mean of 1511 sludge samples from Swiss waste water treatment plants amounts to 410 g organic matter, 73 g Ca and 25 g P per kg dry matter.

Heavy doses of sewage sludge were applied in field and lysimeter experiments in order to assess the influence of sludge on the stability of soil aggretates, on the volume of pores and on the humus content. Only in few cases favorable effects of sludge on these parameters could be observed.

An estimation of the humus balance of the agricultural land of Switzerland showed that about 1 % of the organic matter supply originates from sewage sludge application.

6. REFERENCES

(1) FURRER, O.J. (1977) : Einfluss hoher Gaben an Klärschlamm und Schweinegülle auf Pflanzenertrag und Bodeneigenschaften. Landw. Forsch. Sh. 33/1, 249-257.

(2) FURRER, O.J. (1979) : Die Wirkung von Klärschlamm und Müll-kompost auf Pflanzen, Boden und Sickerwasser in einem Lysimeterversuch. Bodenkundliche Gesellschaft der Schweiz, Bulletin Nr. 3, 73-82.

(3) FURRER, O.J. (1980) : Landwirtschaftlicher Wert des Klär-schlammes. EAS-Seminar "Landwirtschaftliche Verwertung von Abwasserschlämmen" Basel, 24.-26. Sept. 1980, 4.4/1-4.4/11.

(4) FURRER, O.J. (1981) : Accumulation and leaching of phospho-rus as influenced by sludge application. Proceedings on the EEC-Seminar, Groningen NL, June 12 - 13, 1980 on "Phosphorus in Sewage Sludge and Animal Waste Slurries". 1981 : 235-240.

(5) FURRER, O.J. (1981) : Siedlungskomposte in der Schweiz - positive und negative Aspekte. Vortragstagung "Siedlungs-abfall-Verwertung und Nahrungsqualität" 26.3.1981, Speyer.

(6) FURRER, O.J. and BOLLIGER, R. (1981) : Phosphorus Content of Sludge from Swiss Sewage Treatment Plants. Proceedings on the EEC-Seminar, Groningen NL, June 12 - 13, 1980 on "Phosphorus in Sewage Sludge and Animal Waste Slurries". 1981 : 91-98.

(7) SIEGENTHALER, A. and GUPTA, S.K. (1981) : Düngerwert von Klärschlamm. Vortrag LBL, 13.5.1981.

DISCUSSION

Dr CATROUX :   A comment.

With a recommendation of a 2.5 t D.M. sludge to be spended per ha and per year, we are coming back to my Vienna Symposium proposal : sludges may be spread with a spoon !

Dr FURRER :   The Swiss Sewage Sludge Ordinance limit the applicaton rate to 7.5 t DM/ha in 3 years. Liquid sludges is normally applied at a rate of 50 m^3/ha (= 2.5 t DM for sludge containing 5 % DM). 50 m^3 are also a normal dose for cow slurry with normal 3-4 % DM.

Dr DIEZ :   2.5 t per ha and year are a quite common amount of sludge application, i.c. 50 m^3 liquid sludge with 5 % dry matter, about the amount of liquid manure (gülle) the farmers are used to applicate.

Dr WILLIAMS :   Would you really expect any significant differences in stability when dealing with soils of high inherent stability ?

Dr FURRER :   No, not measurables one with normal reasonable doses.

Dr DE HAAN :   You applied, in your experiments, amounts of P comparable with amounts applied in practice in 3 years or more. Do you think that P applied with large amount of sludge will satisfy the crop needs for P for such a long period or will it become unavailable in soil ?

Dr FURRER :   The availability of phosphorus of different sludges can be rather different. In our pot and field experiments we found a rather good P availability in most of the sludges, but not

so good as in superphosphate. Over a long
period, we think the difference from superphosphate will be less (effect of transformation in the soil).

INFLUENCE OF INCREASING AMOUNTS OF SEWAGE SLUDGE ON

THE SOIL STRUCTURE

H. BORCHERT

1. INTRODUCTION

In this paper it is reported about results of soil physical investigations which have been obtained at the Bavarian State Institute for Soil and Plant Cultivation within the scope of the research program "Hygienization of sewage sludge". The influence of differently high amounts of sewage sludge (130, 400 and 800 m^3/ha) with different hygienizing treatments (pasteurization, irradiation) on soil structure should be followed.

2. MATERIAL AND METHODS

During 1973 to 1978 following of sewage sludge were applied :
130 m^3/ha/year; 400 and 800 m^3/ha triennially. If possible the investigations were carried out every spring and autumn. A control plot was compared with the variants with sewage sludge. This plot was situated in the centre of the trial terrain, but less influenced by soil treatments in consequence of continuously lower amounts of sewage sludge in the adjacent plots.

There were investigated :
Distribution of size particles, of volume of pores, and aggregates; water and air permeability; water content; pH value; $CaCO_3$ content; exchange capacity, and cover with Ca, Mg, K, Na-ions and content of Cl-ions.

The trials were carried out on 4 different soils with the same crop rotation : maize, winter wheat, and summer barley. The first location Puch-Straßfeld is situated 550 m over sea level with an annual precipitation of 856mm. The soil is a very loamy silt, and a gley-like Gray Brown Podzolic from Loess. The high content of silt in the upper soil causes an unstable soil structure.

The second location Puch-Neuriß is situated on Würm gravel 540 m over sea level, and presents stones until the mould as an flat plant location. The soil is a sandy loam, and soiltypologically a flat loosening brown earth.

The third location is Neuhof, 518 m over sea level, with an annual precipitation of 764mm. The soil is a gley-like soil from a loamy Jura's covery. The content of clay which increases suddenly in a deep of 70 cm has a water stagnation effect.

As last location was choosen a sandy soil, namely Baumannshof, 365 m over sea level, with an annual precipitation of 700 mm. Soil-typologically a gley with a sinked groundwater, this soil was created from humous until completely alluvial sands.

3. RESULTS

Tables 1 to 5 show only soil values which during the investigation period demonstrated significantly the influence of different amounts of sewage sludge applied on the correspondent soil, namely : volume of pores, content of air pores, and utile field capacity, furthermore aggregate formation, exchange capacity, and content of Ca and Mg in the sorption complex. In table 1 values are shown of the total volume of pores. The real values of all investigation years were taken for spring and autumn. These average values are stated as relative numbers for control which is = 100. (Table 1).

TABLE 1 : CHANGE OF TOTAL-PORE-VOLUME (VOL. %) AFTER APPLICATION OF SEWAGE SLUDGE

	Control	130 m³			400 m³			800 m³		
	= 100	un-treated	pasteur-ized	irrad-iated	un-treated	pasteur-ized	irrad-iated	un-treated	pasteur-ized	irrad-iated
1974 – 1978										
Straßfeld spring	46.6	96	95	100	95	98	93	97	98	98
autumn	44.9	98	101	97	99	102	98	98	101	100
1974										
Neuriß spring	48.7	105	102	98	98	95	93	104	97	101
autumn	51.5	103	99	97	99	96	101	91	98	97
1974 – 1978										
Neuhof spring	43.3	103	103	105	104	103	104	103	102	106
autumn	41.5	100	100	101	99	101	101	100	103	101
1974, 1975, 1977, 1978										
Baumannshof spr.	51.9	105	101	103	101	103	100	101	99	105
autumn	45.6	104	102	101	103	99	96	103	108	110

It is remarkable that already the control values are nearly always higher in spring than in autumn. In Straßfeld, application of sewage sludge did not cause an increase of total volume of pores, in spring more a compression. In Neuhof (second loess location), application of organic matter caused in all plots an increase of total volume of pores, in spring more significantly. Reason for that could be the darmming up of water, the organic substance decomposes slower. In Baumannshof, application of sewage sludge increases nearly always the total vacuum, especially by application of 800 m^3/ha in autumn.

The location Neuriß with a soil rich of vacuums is little influenced, and no shows clear tendencies.

Important is the indication of total volume of pores, more important the distribution of size of pores, specially the changes in the part of air-pores. (Table 2) Springing of values and the little uniform tendency in the location of Neuriß are clearly shown, because of the stoniness of the soil no further pore investigations were carried out, except in 1977. In the sandy soil of Baumannshof the content of air increased apparently by application of sludge. From the two loess soils, the more silty soil of Straßfeld shows a decrease of content in air, specially in the spring values. The more loamy soil of Neuhof was porouser by application of humus, the content of air increased. In the variants with 130 m^3/ha seems to appear additionally an accumulation effect by annual application of sludge. The content of air decreased again in autumn.

One of the most important soil values is the utile field capacity: plant-available water (Table 3) in Neuriß, the pores of the utile field capacity increased always apparently by application of 800 m^3/ha. By application of 130 and 400 m^3/ha the values no show clear tendency. The sandy soil of Baumannshof was clearly more capable to absorb plant - available water by application of sludge. Certainly, the high values by application of 130 m^3/ha sludge appeared with the accumulation effect. In Straßfeld, the application of sludge influenced hardly the content of pores

TABLE 2 : CHANGE OF AIR-CONTENT (VOL. %) AFTER APPLICATION OF SEWAGE SLUGE (PORES > 50 µ)

	Control	130 m³			400 m³			800 m³		
	= 100	un-treated	pasteur-ized	irrad-iated	un-treated	pasteur-ized	irrad-iated	un-treated	pasteur-ized	irrad-iated
1974 – 1978										
Straßfeld spring	10.9	77	66	97	66	88	57	83	79	85
autumn	8.9	108	112	84	75	103	86	80	93	86
1974										
Neuriß spring	15.4	129	118	87	99	72	73	112	70	82
autumn	15.7	117	82	92	90	97	112	73	123	106
1974 – 1978										
Neuhof spring	5.6	126	160	108	153	146	123	94	112	137
autumn	5.8	96	103	105	108	96	127	86	94	93
1974, 1975, 1977, 1978										
Baumannshof spring	26.9	82	82	97	80	85	89	69	59	81
autumn	14.9	63	69	77	84	81	88	87	96	100

TABLE 3 : CHANGE OF UTILE CAPACITY OF WATER (VOL. %) AFTER APPLICATION OF SEWAGE SLUDGE

(pores 50 - 0.2 µ)

	Control	130 m²			400 m²			800 m²		
	= 100	un-treated	pasteur-ized	irrad-iated	un-treated	pasteur-ized	irrad-iated	un-treated	pasteur-ized	irrad-iated
1974 – 1978										
Straßfeld spring	22.2	98	98	100	104	96	101	104	106	103
autumn	21.8	93	91	87	100	100	101	98	100	104
1974										
Neuriß spring	21.1	90	93	102	100	103	91	111	111	120
autumn	18.0	112	133	100	128	79	76	132	108	117
1974 – 1978										
Neuhof spring	19.6	111	109	108	117	115	111	120	113	120
autumn	20.2	99	98	102	97	98	93	96	100	98
1974,1975, 1977,1978										
Baumannshof spring	19.9	115	112	105	116	113	107	124	106	135
autumn	23.0	116	115	113	99	102	98	104	104	107

with plant-available water, except an insignificant increase by application of 800 m^3/ha. In Neuhof the application of humus conducted to an increased field capacity in the spring values corresponding to the increased amounts of humus. But this effect decreases significantly during the vegetation period.

As an expression of change in the structure are stated in table 4 the values of contents on great soil aggregates after an immersion sifting. The values of the investigation year 1977, nearly on the end of the experiment period, were drawn out. The low values of the control in the locations of Straßfeld and Baumannshof in spring are conditioned by the high content of silt and sand as well as by the preceding crop (summer barley) which makes worse the structure, while the maize which develops the structure increases the values in 1977. Compared with the initial values of the control, application of sludge conducted to very high aggregate values during spring in the locations of Straßfeld and Baumannshof. Also in Neuhof, and by application of 800 m^3/ha in Neuriß, the content of great aggregates increased. But during the year this additional structure - favouring effect of sludge application decreased in all locations with all variants.

This structure - favouring effect of sludge application is based on supply of fine material, specially such of organic origin. The exchange capacity increases as shown in table 5. Ca and Mg content of the sludge seems to have - shortly and by no too high precipitations (danger of erosion) - a structure-favouring effect with formation of greater aggregates and a more favourable porosity in the soils treated with sludge.

By application of sludge in the field, specially by application of 800 m^3/ha, a more quickly seeping and drying of the variant "irradiated" appeared. The soil structure values which were found are higher by the variant "800 - irradiated" in the most cases compared with the control, and equal or higher compared with "pasteurized". The scissure formation after drying of sludge with 105° C shows a lower shrinking by "untreated" and "pasteurized" than by "irradiated" sludge. Soil physics cannot give an explication for these observations.

TABLE 4 : CHANGE OF NUMBER OF AGGREGATES AFTER APPLICATION OF SEWAGE SLUDGE

(% aggregates 6-2 mm beginning with 6-5 mm aggregates)

	Control = 100	130 m³			400 m³			800 m³		
		un-treated	pasteur-ized	irrad-iated	un-treated	pasteur-ized	irrad-iated	un-treated	pasteur-ized	irrad-iated
1977 Straßfeld march	52	175	173	171	146	113	160	156	148	138
october	97	99	99	99	95	100	101	99	100	98
1977 Neuriß march	88	108	98	98	84	93	86	100	106	109
october	92	100	104	100	103	103	105	103	105	105
1977 Neuhof march	93	103	102	103	105	106	106	106	90	102
october	97	100	93	100	99	99	101	101	100	100
1977 Baumannshof march	61	131	123	113	118	118	100	123	133	121
october	86	106	101	103	101	102	97	106	107	102

TABLE 5 : CHANGE OF EXCHANGE CAPACITY (a), of Ca-SORPTION (b), AND OF Mg-SORPTION (c)

IN MVAL/100 G SOIL

		Control	130 m³			400 m³			800 m³		
		= 100	un-treated	pasteur-ized	irrad-iated	un-treated	pasteur-ized	irrad-iated	un-tread	pasteur-ized	irrad-iated
Straßfeld	a)	12.8	103	105	103	98	101	98	109	108	108
1978	b)	8.9	108	110	106	94	102	101	112	111	112
	c)	0.52	179	169	150	135	140	154	210	200	204
Neuriß	a)	19.2	97	96	96	99	103	98	105	105	102
1978	b)	15.0	90	89	91	96	103	99	107	107	104
	c)	2.38	105	96	100	103	106	110	113	111	109
Neuhof	a)	12.6	96	102	102	106	106	103	110	106	108
1978	b)	10.8	99	103	100	102	104	102	106	106	102
	c)	0.47	104	119	121	132	132	117	162	138	136
Baumannshof	a)	12.4	110	103	85	98	83	80	121	128	124
1978	b)	3.9	185	128	118	169	131	113	236	272	251
	c)	0.31	252	171	139	197	155	139	297	303	287

4. CONCLUSIONS

The effect of sewage sludge on soil was more significant in spring than in autumn. In physical respects, the influence of sewage sludge on the soil structure is insignificant, specially by amounts of sewage sludge for practical purposes, favourable by sandy soils, and temperary unfavourable by silty soils. The nutrient effect influences more soil and plant.

5. SUMMARY

Effect of sewage sludge on soil was more significant in spring than in autumn. In physical respects, the influence of sewage sludge on soil structure is insignificant especially by amounts of sewage sludge for practical purposes, favourable by sandy soils, and temperary unfavourable by silty soils. Nutrient effect may more influence soil and plant.

THE INFLUENCE OF THE AGRICULTURAL UTILIZATION OF

DOMESTIC SEWAGE SLUDGE ON THE QUALITY OF THE SOIL

J. BORTLISZ AND F. MALZ

1. The agricultural use of the digested domestic sludge is of
some general economic importance as far as the utilization
of the sludge and the minimizing of the ultimate elimination
costs are concerned, as well as on the background of an increase in prices of imported mineral fertilizers. Yet this
striving for cutiing down the expenses may not lead to a
neglect of the necessary precautions, which are crucial to
the utilization of domestic sludge. This is especially true
for the nature of the ground and the avoidance of any in-
jury to human or animal health.

The influence of digested domestic sludge on the soil was
one of many aspects examined by the Lippeverband Essen,
being in close contact with the Josef-König-Institut Münster,
in a test series running for ten years.

2. The field on which the experiments were carried out consisted
of forty lots each having an acreage of 25 m^2. Their arrange-
ment and the structure of the soil is displayed by Picture 1.
The soil consists of loamy or very loamy sand and can thus
be seen as typical of that region where the sludge is being
produced and where it is to be utilized.

During the first period of the test series, lasting for five
years, we examined whether there are any differences between
mulching the lots either with sewage sludge or with manure,
and if so, which exactly these differences are. Therefore
the lots were dunged with manure and sewage sludge, respectively, thrice during this first period (Picture 2).

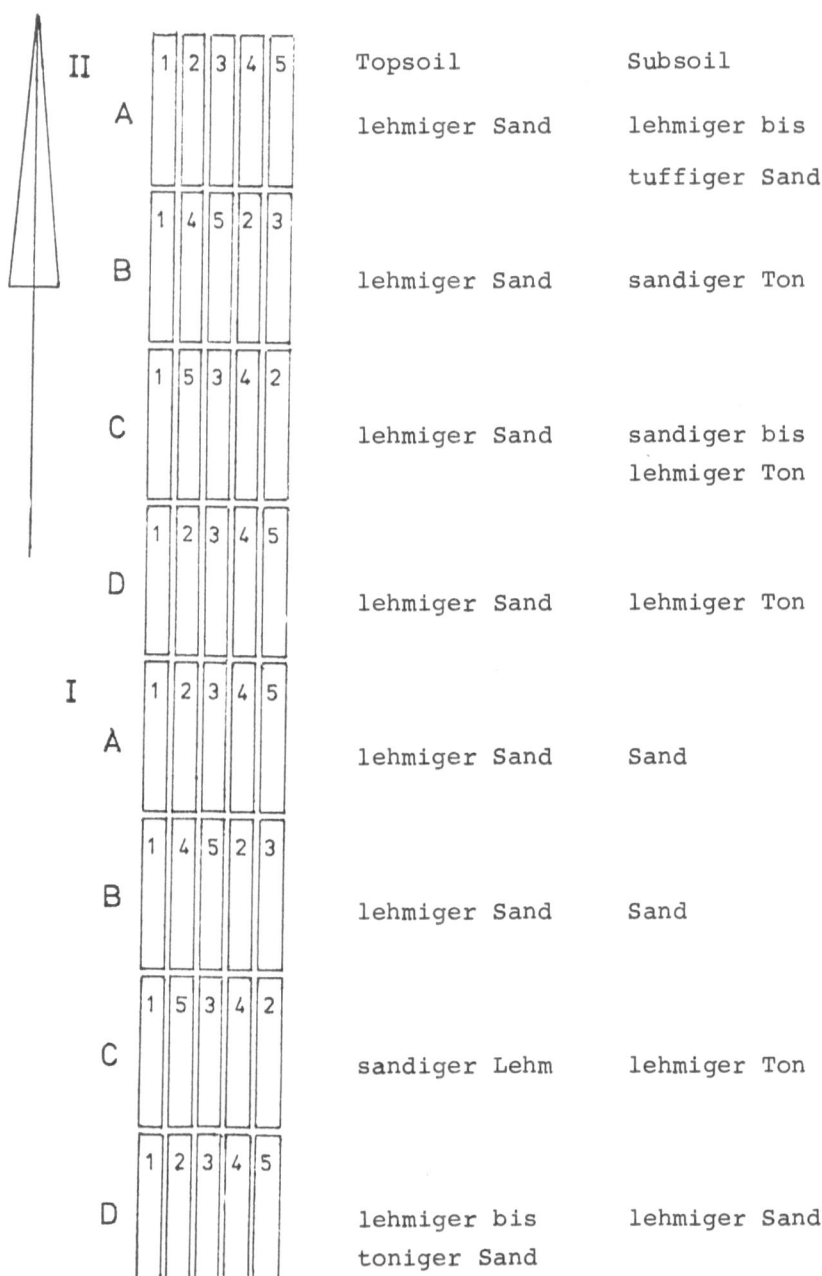

	Topsoil	Subsoil
II A	lehmiger Sand	lehmiger bis tuffiger Sand
B	lehmiger Sand	sandiger Ton
C	lehmiger Sand	sandiger bis lehmiger Ton
D	lehmiger Sand	lehmiger Ton
I A	lehmiger Sand	Sand
B	lehmiger Sand	Sand
C	sandiger Lehm	lehmiger Ton
D	lehmiger bis toniger Sand	lehmiger Sand

M = 1:500

Picture 1 : Testallotments / kind of soil

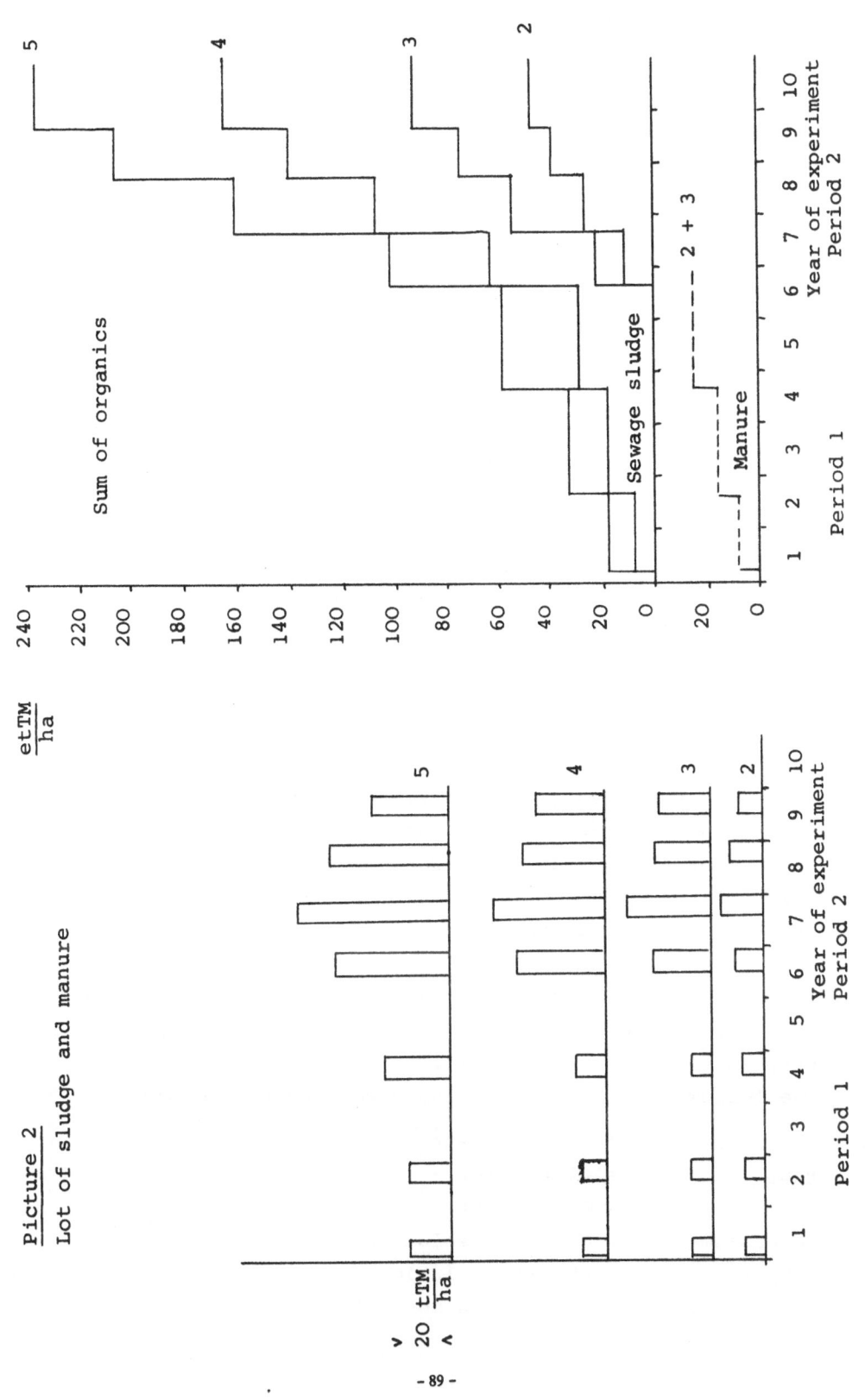

Picture 2
Lot of sludge and manure

Sum of organics

During the second five-year-period the soil was treated with too high a quantity of sludge, in order to simulate the influence of high doses of sludge on the soil and on the crop (Picture 2).

At the same time the comparison lots 1 were treated with mineral fertilizers only, the lots 2 received a small amount of mineral compensation, while the lots 3, 4 and 5 merely obtained a potash compensation.

Digested domestic sludge drained in the drying bed was being used for the experiments. The drainage area of the sewage plant was not subject to any noteworthy or essential industrial load. The dry sludge made an equal dispersion and a gradation of the doses possible. Immediately after being applied the sewage sludge or the stable manure, respectively, were skilfully mixed with the soil.

Table 1 shows the average values of the data of the sludge and the manure.

Different types of grain and field grass were tested in the experiments. The rotation of crops is shown by Table 2. The produce of every single lot was determined, and the average value of the data of the repetitions, i.e. the parallel deposits of every variant, was taken.

3. The soil analyses were made between the harvest and the new fertilizing. A layer of 25 cm was determined to distinguish between top-soil and subsoil.
   The following analytical methods were being applied :

   pH (KCl)   = electro-metric
   P_2O_5 = DL-extract
   K_2O = DL-extract
   magnesium   = $CaCl_2$-dissolution
   copper   = HNO_3-dissolution
   boric acid   = soluble in hot water

Table 1 : Characteristics of sludge and manure

| | | | Sludge | | Manure |
			1. Period	2. Period	1. period
Drymass	TM	%	40	47.5	34.8
Chlorid	[Cl]	g/kg TM	0,7	0,5	
Glowloss		"	366	314	
Org. Carbon	[C]	"	221	185	134
Nitrogen (Kjeldahl)	[N]	"	16	16	6.5
Phosphor	$[P_2O_5]$	"	56	58	7.8
Lime	[CaO]	"	76	56	
Potash	$[K_2O]$	"	3.6	3.1	6.3
Magnesium	[MgO]	"	16	13	3.6
Iron	[Fe]	"		20	
Zinc	[Zn]	mg/kg TM	3016	3042	
Mangan	[Mn]	"	283	477	
Copper	[Cu]	"	262	212	20
Lead	[Pb]	"	93	203	
Chromium	[Cr]	"	70	76	
Nickel	[Ni]	"	64	49	
Cadmium	[Cd]	"		9	
Cobalt	[Co]	"		18	
Mercury	[Hg]	"		4	

Table 2 : Rotation of crops

Year	Field I	Harvest	Field II	Harvest
1	Summer rye	14.8.	Summer barley	14.8
1/2	Winter rye	10.8.	Winter rye	10.8
2/3	Fieldgrass	24.5. - 27.9.	Fieldgrass	24.5 - 27.9.
3/4	Winter rye	18./22.8.	Winter rye	18./22.8.
4/5	Oats	13.10	Winter rye	13.10
5/6	Winter barley	25.7.	Barley/Oats	8.8
6/7	Fieldgrass	2.6. - 24.7.	Fieldgrass	2.6. - 24.7.
7/8	Winter rye	26.7.	Winter rye	26.7.
8/9	Winter barley	3.8.	Winter barley	3.8.
9/10	Fieldgrass	1.6. - 10.8.	Fieldgrass	1.6. - 10.8.

zinc   = HCl-dissolution
N   = Kjeldahl-dissolution
humus (org. C)   = incineration at 450°C (first period);
                          determination of the org. C by means
                          of carboanalyzer (second period).

4. The dry substance of the digested sludge consists of
   32-38 % of organic material, given a conversion of 1,72 x
   organic carbon.

The experiments show, dependent upon the quantity of sewage
sludge applied, that the amount of humus within the top-soil,
which is the main area of utilization, increases, whereas
in the subsoil no changes can be observed (Picture 3).

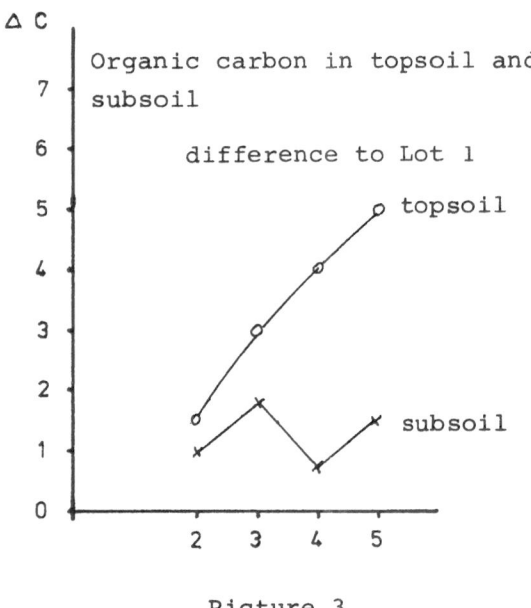

Picture 3

Only a negligible quantity of organic carbon is transported vertically   Under these circumstances, one can conclude from the ratio between the amount of carbon supplied
and the increase measured in the soil (Table 3), that the
speed of the transformation of the organic mass in the

soil grows with increasing quantities of sludge being applied.

TABLE 3 : INCREASE OF ORGANIC CARBON IN TOPSOIL

(weight of topsoil = 4000 t dry substance/ha).

Lot	Supply g C/kg Soil	Increase g C/kg Soil	Supply / Increase
2	2	1.5	1.3
3	4	3	1.3
4	6	4	1.5
5	8	5	1.6

5. The sewage sludge contains high quantities of components
   having a basic effect, like calcium and magnesium, which
   results in a most desirable stabilization of the pH-value
   or a shift to higher pH-values, respectively. Picture 4
   shows the relative pH-alterations of the soil as compared
   to the zero-lot.

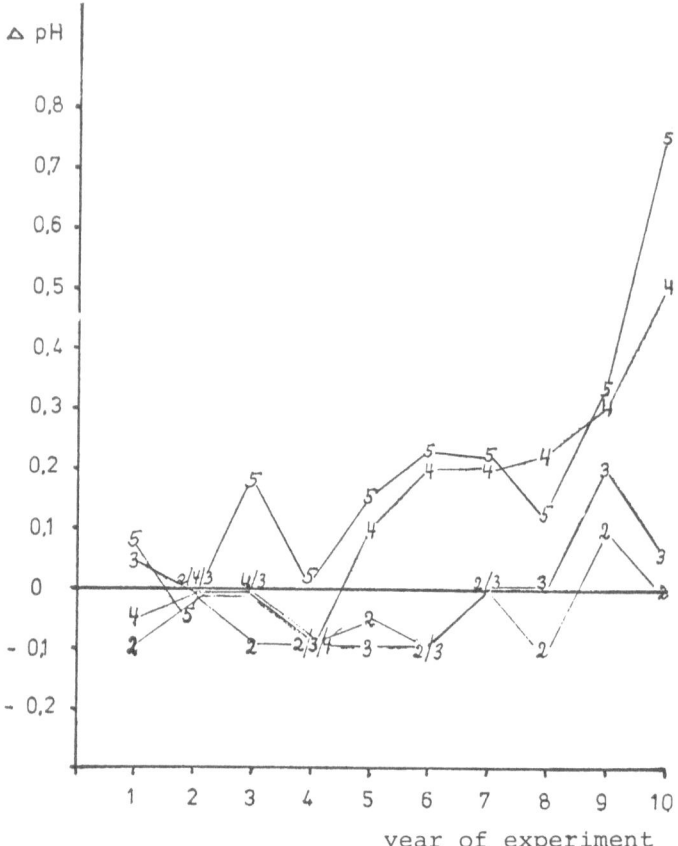

Picture 4 : p_H in topsoil, difference to Lot 1.

Besides the stabilizing effect of small quantities, the advantageous pH-alterations into the alkaline range become evident, when higher doses are applied. This pH-stabilizing or -alkalizing effect of the sludge is especially important for the fixation and immobilization of heavy metals in the soil with regard to the prevention of their assimilation by the crop as well as their vertical transport in the subsoil up to the ground water.

6. The phosphor of the domestic sewage sludge is a most valuable component within the scope of agricultural utilization. In our experiments we examined the behaviour of the plant-available phosphor in the top-soil and the subsoil considering its dependence on the quantity of sludge being applied. As shown above, we applied relatively small doses of sludge in the first test-period. Thus in a period of 5 years, the lot 4 was mulched with 5 t and the lot 5 with 10 t of dry matter per hectare and annum. Only the latter lot showed a small increase of the amount of phosphor measured in the top-soil (Picture 5).

No lot, though, not even lot 5, showed an increase of phosphor in the subsoil (Picture 6). Only the extremely increased quantities of sewage sludge which were applied in the second test-period led to growing amounts of phosphor in the top-soil, in relation to the doses the respective lots were treated with. These deliberately excessive quantities led to small increases of the amount of phosphor in the subsoil at the end of the second test-period. A valuation of the inter-action between the amount of phosphor being supplied, the abstraction by the crop according to average agricultural data and the measured plant-available P_2O_5-concentration of the soil shows, that the relative share of not immediately plant-available phospher grows with increasing quantities of sewage sludge being applied. (Table 4)

TABLE 4 : P_2O_5-RELATION BETWEEN SUPPLY, DEPRIVE BY CROPS AND AMOUNT IN TOPSOIL

Lot	$\dfrac{\text{[Supply]} - \text{[Deprive]}}{\text{[Available]}}$
1	1,5
2	1,5
3	2,2
4	2,5
5	2,7

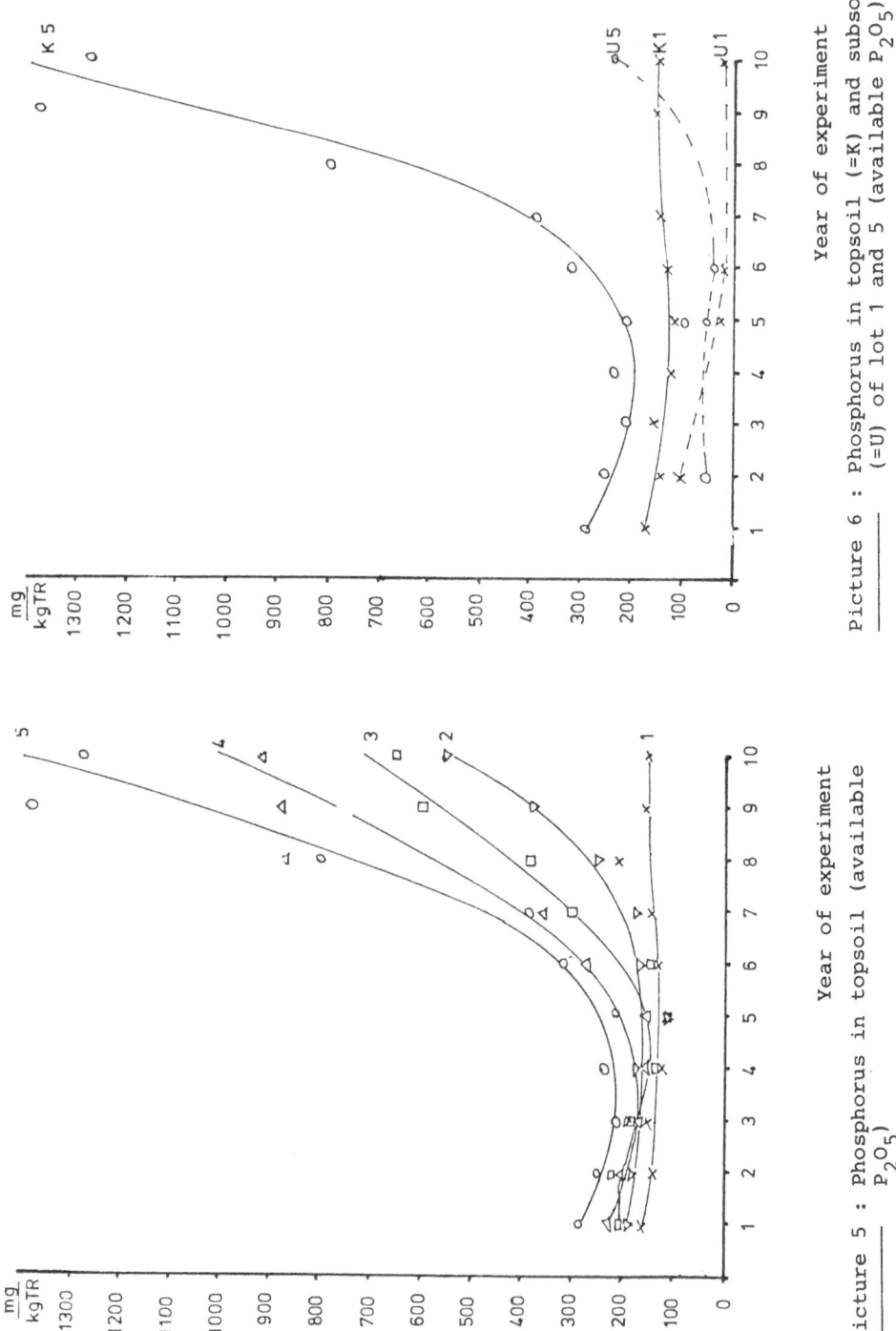

Picture 5 : Phosphorus in topsoil (available P_2O_5)

Picture 6 : Phosphorus in topsoil (=K) and subsoil (=U) of lot 1 and 5 (available P_2O_5)

It may thus be concluded, that high amounts of sludge provide the soil with a longer lasting supply of available phosphor, while at the same time no shifts of the pH-value to lower values or other negative alterations of the soil will occur. Furthermore, some part of the phosphor within the sludge seems to be fixed more steadily. During a long-term fertilizing with sewage sludge the P_2O_5-concentration of the soil should therefore also be observed.

Given the soil structure of the test-lots, i.e. loamy to very loamy sand, a sludge amount of 5 to 10 t of dry mass per hectare and annum seems to be the optimal quantity for keeping a good P_2O_5-level in the soil; under these circumstances there is no need for adding any mineral phosphor-fertilizer.

7. On the background of the topical discussions about heavy metals in connection with sewage sludge fertilizing, this subject will be touched now. It was one of the main issues of our investigations.

As compared to other heavy metals, there is normally a relatively high zinc concentration within domestic sewage sludges (Table 1). The sludge applied to our lots contained approximately 3.000 mg of zinc per ka of dry mass.

The comparison between the zero-lots and those treated with the highest amount of sludge (Picture 7) showed, that - as we expected - the zinc-concentration in the top-soil increased during the test-period in accordance with the quantity of sludge being applied. Yet the result, that in spite of the enrichment of the top-soil only a very small vertical transport into the subsoil could be observed, is of crucial importance (Picture 7/8).

Picture 9 shows the heavy metal concentrations which were found in the top-soil at the end of the test-period, in relation to the limiting values for soil. It can easily be seen

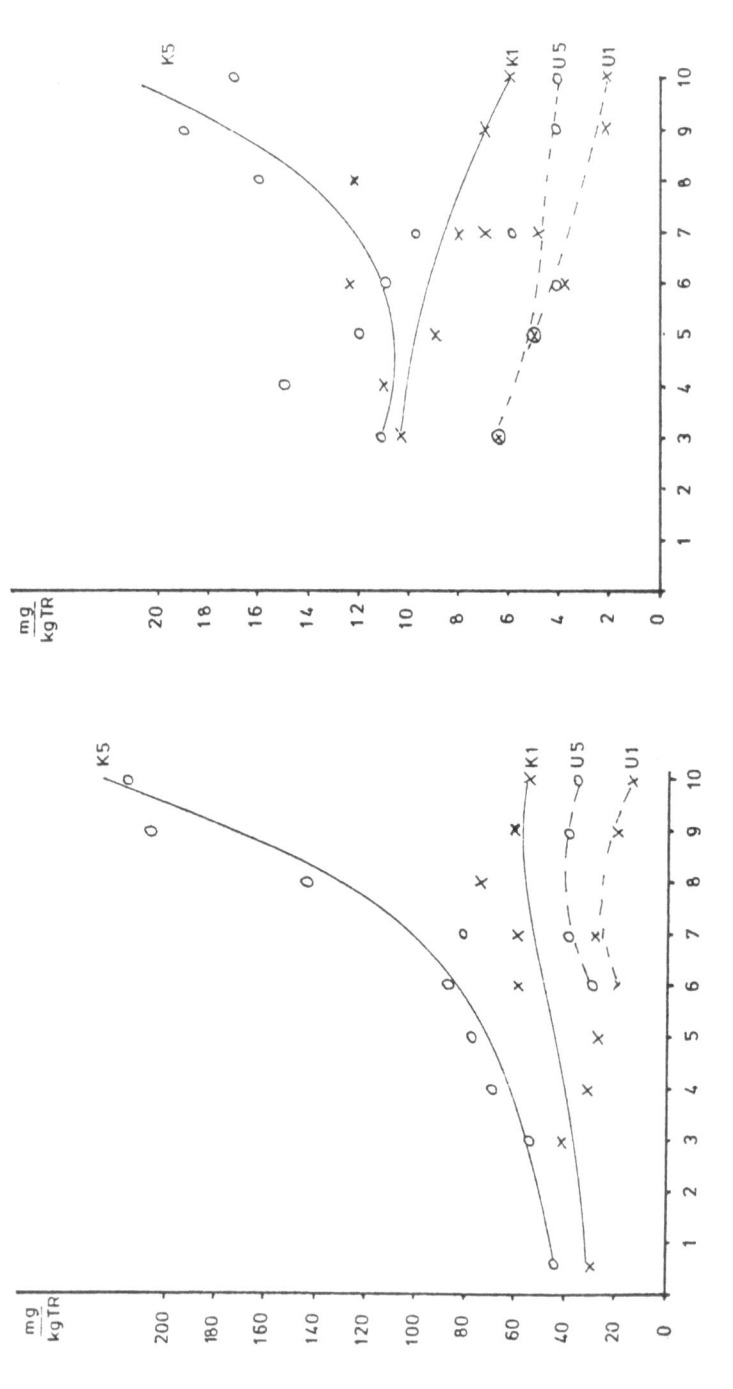

Year of experiment

Picture 7 : Zinc in topsoil (=K) and subsoil
(=U) lot 1 and lot 5.

Year of experiment

Picture 8 : Copper in topsoil (=K) and subsoil (=U)
lot 1 and lot 5.

that its growth is dependent on the quantity of the dose being excessively increased with a fast motion effect; nevertheless, there was no transport into the subsoil.

The charging regulations determine, among others, limiting values for heavy metals occuring in the soil. Table 5 (column 5 + 6) shows the waiting period for each heavy metal on the basis of a sewage sludge fertilizing of 5 t of dry mass per heactare and annum. At any rate, a qualtifying discussion about the crucial concentration of the omnipresent metal zinc has to take place, in order to determine new limiting values and to derive conclusions for a long-term utilization of the sewage sludge without straining the ecological conditions.

8. The productivity of the soil is an essential qualitative feature for its utilization. Another main issue of our experiments was the influence of the sewage sludge on the productivity of the soil. The rotation of crops was already displayed by Table 4. The application of sewage sludge led to a significant increase of the day mass output of the grain as well as of the field grass.

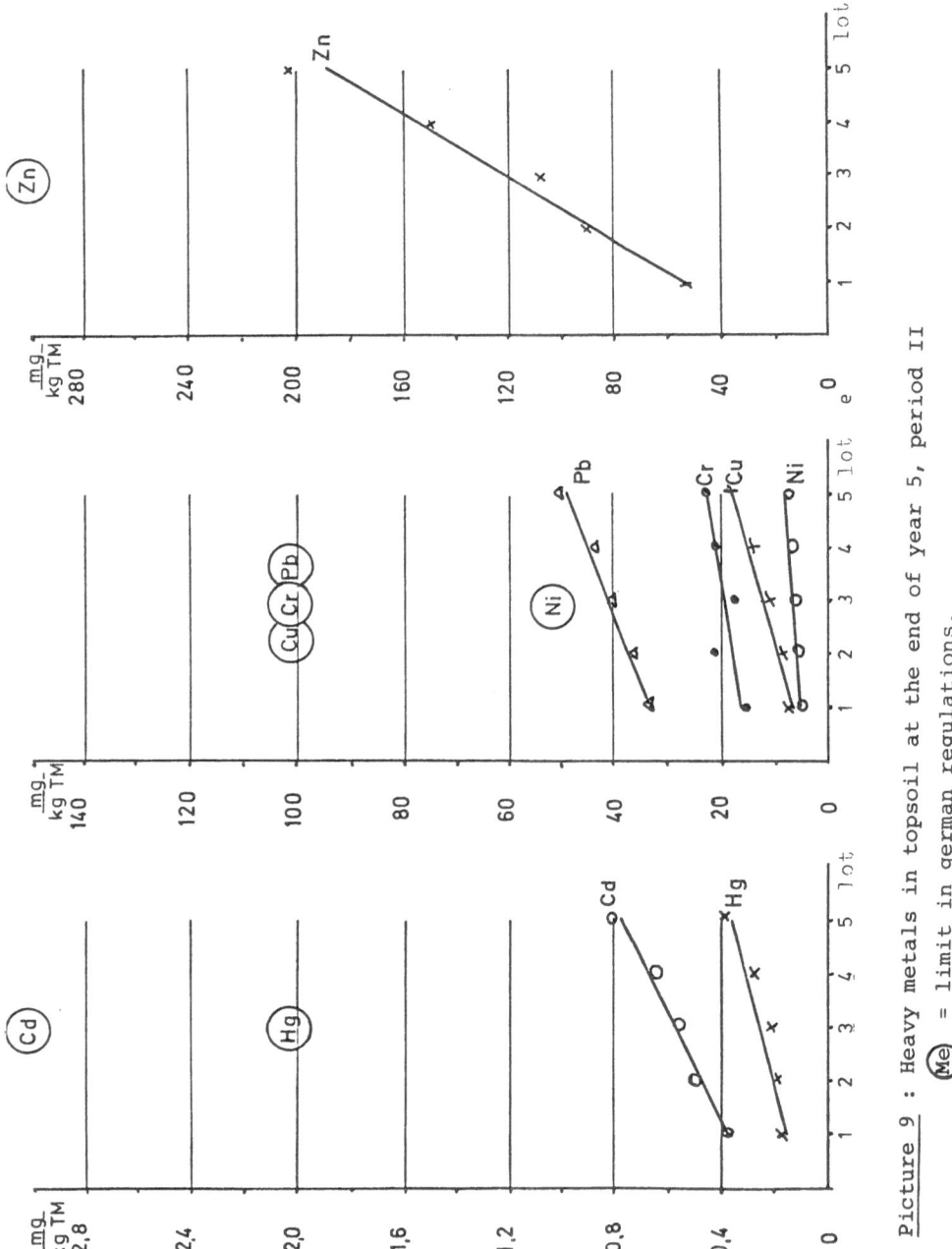

Picture 9 : Heavy metals in topsoil at the end of year 5, period II

(Me) = limit in german regulations.

TABLE 5 : **HEAVY METALS IN TOPSOIL, LIMIT VALUES (GERMAN REGULATIONS) IN OPPOSITION TO RESULTS OF EXPERIMENTS OF THE LIPPEVERBAND**

Element	Limit value (§ 15 Abfg.) mg/kg TR	Amount in topsoil (beginning) mg/kg TR	Average values of the sludge mg/kg TR	Years to reach the limit values in soil in relation to	
				Analysis of Sludge a	Analysis of Topsoil b
Mercury	2	0,17	4	315	230
Cadmium	3	0,4	9	202	163
Nickel	50	5	49	642	448
Copper	100	8	212	303	204
Chromium	100	17	76	764	334
Lead	100	36	203	220	104
Zinc	300	60	3042	55	43

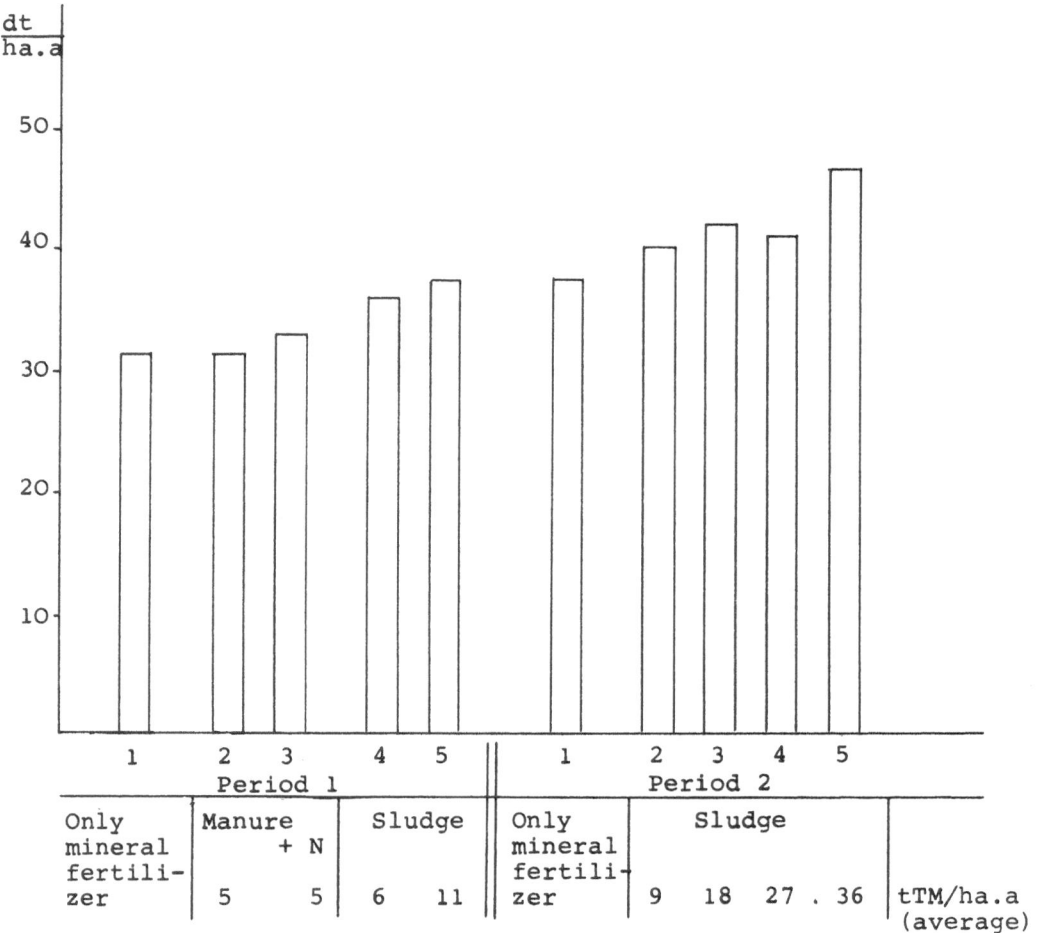

Picture 10 : Average output (corn dry mass)

In picture 10 is shown the output (corn drymass) of period 1 : relation between manure (lot 2,3) and sludge (lot 4,5) and of period 2 : influence of the mass of sludge given on the fields.

Picture 11 shows in relation to lot 1 = 1 the output of field grass and corn of both periods. The output is bigger from the lots dunged with sludge even dunged with manure. Also the output increase with the increase of sludge given on the fields.

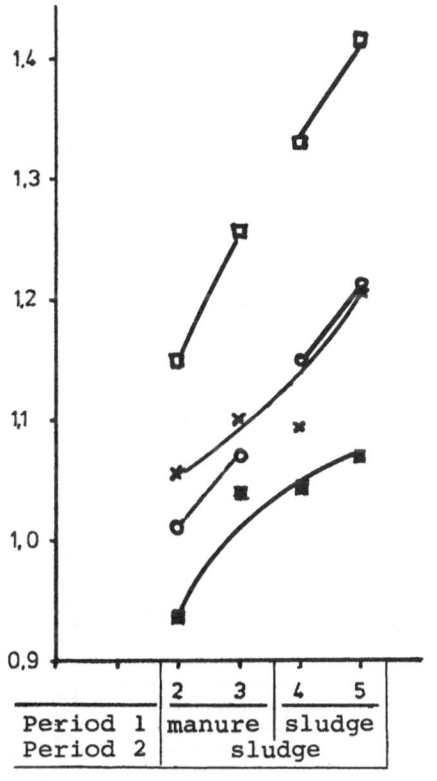

□ = 1. period fieldgrass
o = 1. period corn
■ = 2. period fieldgrass
x = 2. period corn

(lot of sludge and manure see picture 2 or 10)

Picture 11 : Relative output (lot 1 = 1), field grass and corn.

- 103 -

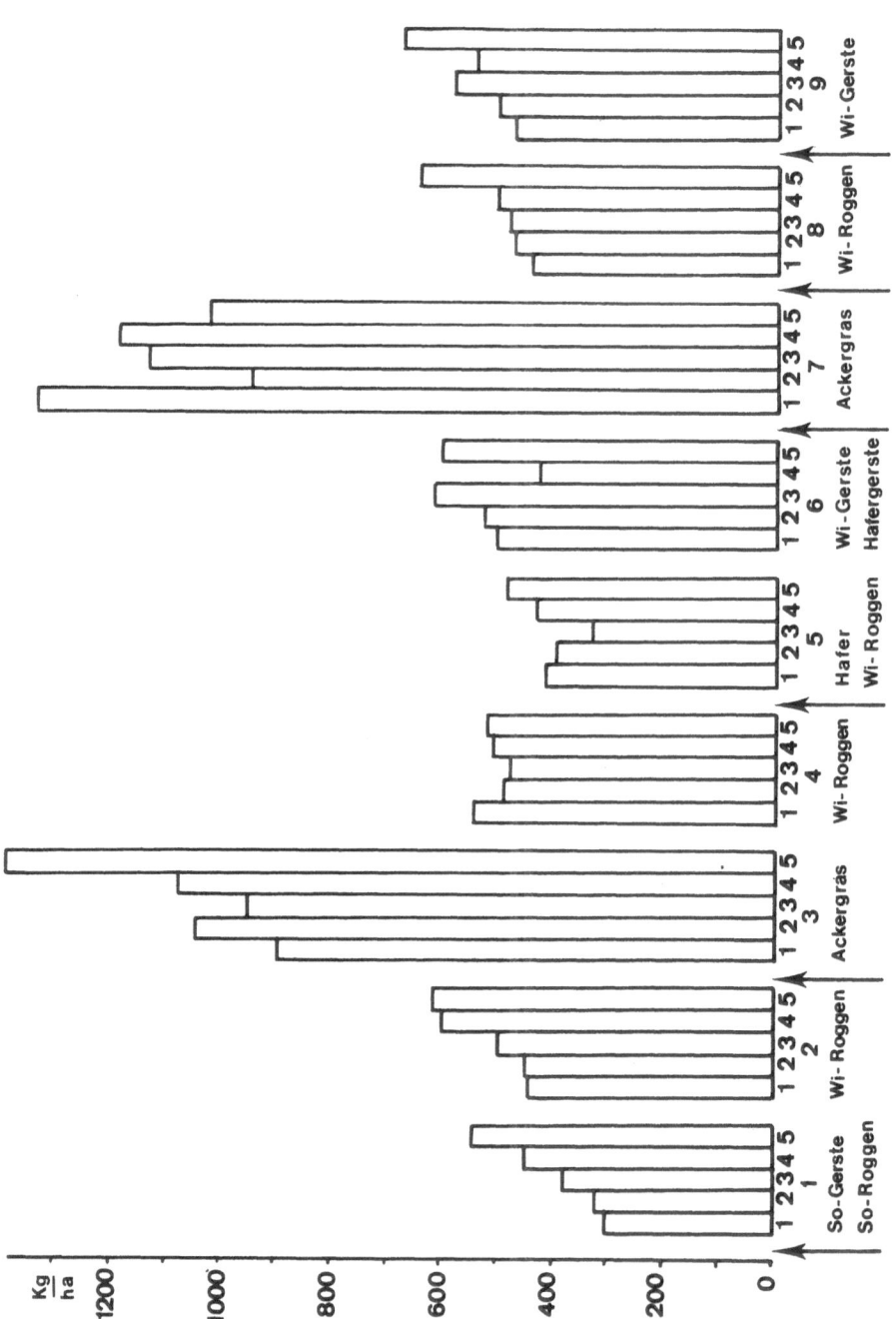

Picture 12: Albumen output.

There was also an increase in the protein quantity being produced (Picture 12).

The sewage sludge can thus be seen as equal to other organic fertilizers, like e.g. stable manure. To make the most of its fertilizing properties, the quantity added at once should contain 5-10 t TM/ha.a according to the results of our experiments. Yet the charging regulations determine a maximum amount of 7,5 t TM/ha within a period of three years or 2,5 t TM/ha.a, respectively. But it is nevertheless possible to optimize the utilization of the sewage sludge and the output of the crop, if the sludge is applied at once and a mineral compensation is added in the intervening years, regarding the nitrogen-concentration and the decaying effect of the nutritive materials of the sludge.

If dry bed sludge containing only a small nitrogen-concentration is applied, there is normally no danger of overdosed fertilizing. But attention has to be paid, if sewage sludges containing a high nitrogen-concentration are applied; these are digested sludges that have not gone through intermediate storing and have not been drained mechanically.

9. Our series of experiments, running for ten years, has proved that the digested domestic sludge is a rich nutritive organic fertilizer which is of the same value as manure. The application of sewage sludge provided the top-soil with a good and sufficient supply of humus and nutrients.

For the time being, the concentration of heavy metals is still a limiting factor. But the reduction of the emission of heavy metals is a problem that has to be solved by the companies and industries of the respective area, in order to enable the utilization of the sewage sludge as a valuable fertilizer.

DISCUSSION

Dr DE HAAN :   How was plant available P determined ? We determine plant-available P as water soluble P and
then the effect of sewage sludges is mostly not
evident.

Dr BORTLISZ :   We have determined the plant-available P by the
double lactat method. We have no experiences
with the water-soluble method.

Dr BORTLISZ :   Dr CATROUX, is their a general method of measurement of fertility of soil ?

Dr CATROUX :   That is a good question. It is difficult to define the soil fertility from a soil microbiology
point of view. However, I think it is possible
to conclude when you consider some specific microbial activities like nitrification, symbiotic
nitrogen fixation, pesticides biodegradation and
eventually total soil microbial biomass which
can respond drastically to biocidal treatments
or may be to supply of sewage soils loaded with
heavy metals.

Dr DE HAAN :   The best criterium for soil fertility is its
yield level over a long period.

Dr SÜSS :   Did you observe a difference between sewage
sludges and cattle manure effects ? On which
soil type did you perform your experiment ?

Dr BORTLISZ :   The trials were performed on a sandy loam for
10 years. Comparing sewage sludge and cattle
manure, there was a better yield effect with
sewage sludge but the difference in yield is not
statistically significant.

PHYSICAL PROPERTIES IN SEWAGE SLUDGE AND SLUDGE

TREATED SOILS

E. VIGERUST

1. MODE OF USING SEWAGE SLUDGE

Considering the effects of sewage sludge it is important to
know how to use the sludge.

In Norway we have three different kinds of utilization.

1.1. Agricultural use

For using the sludge it is dewatered by 18-25% dry matter content. We have in the agriculture a system with either onesided cereal production or cattle production, where gras is the
main crop. Use of sludge is only a reality in connection with
cereal-production.

Every year large areas of mineral soils are brought into cultivation and it will take long time to build up adequate contents
of humus in soil. In some parts of the country soil levelling
with bulldozers has been very common. It is rather difficult to
take care of the top soil and bring it all back after the
operation. The physical properties of this soil will be bad for
many years and the crop production will be very reduced.

In our experiments we have got the best effect of sludge application on these levelled soils.

We have also other areas where application of sludge may lead to better physical properties.

In experiments with sludge application it is very difficult to distinguish between the effect of fertilizer and better physical properties. Farmers'interest for the utilization of sludge is determined by its capacity as a soil conditioner.

1.2. Use of sludge on green areas

In many urban areas in this country the soil consists of just a thin cover. To protect cultivated soil is of importance owing to the fact that it does not comprise more than 3 per cent of the land area. Therefore, the developing has to take place on non-productive areas. Consequently, there is lack of proper growth medium. Well decomposed sludge may be the compensation for utilization on green areas. For this reason it is important to use relatively thick layers of sludge compost.

1.3. Use of sludge as top-layer on finished landfills

Revegetation of landfills is very important. It is, however, difficult to obtain enough material for good quality topsoil. Dewatered sludge has often been brought to landfills with a view to solve problems connected to disposal of sludge. Experiments here have demonstrated that sludge is well fitted as top soil for sowing and planting. We were successful in our experiments by using pure sludge on 20-50 cm layers. The sludge has to be well decomposed and besides, an effective decomposition is supposed to take place within the storage period.

1.4. Necessity of planting on stored sludge

Covering stored sludge with vegetation will increase the evaporation, sludge will dry up quicker and the amount of leaching

water will be considerably reduced. Nutrients will also be
stored   for future use.

Considering the variety of using possibilities we preferably
studied physical properties of pure sludge instead of sludge
and soil mixtures.

2. SEWAGE SLUDGE AS SOIL CONDITIONER

2.1. Application of organic matter

The effect of sludge on soil physical properties is mainly due
to the content of organic matter. On volum basis the sludge
consists of about 80 per cent organic material.

Application of 10 tons sludge as dry matter per hectar will not
increase humus content in the plough layer more than about 0,2
per cent (on weight basis). This indicates that :

1) it will be difficult to get measurable effects from small
   applications of sludge;

2) the best effect on soil physical properties will be obtained
   on soils low in humus.

On the other hand the effect will last for several years. The
organic material in sludge will gradually be reduced. Our
temporary results indicate that approximately 60 per cent of
organic matter in raw sludge will rapidly decrease to 40-45 per
cent. In the next 2-4 years a further decrease to 25-30 per cent
will occur. The remainder seems, however, to be relatively
stable.

2.2. Application of sludge on levelled soils

NJØS (1978) has carried out two experiments with high application of sewage sludge combined with nitrogen fertilizer on
levelled soils. The crops were oats and barley and the average
results as kg grain/hectar were :

Treatment	d.m.	Exp. 1 1971-73	Exp. 2 1973-77
0 tons sewage sludge m/hectar		2780	1040
150 tons sewage sludge m/hectar		3910	1450
300 tons sewage sludge m/hectar		4540	2110
50 kg N/hectar		3720	1270
100 kg N/hectar		3800	1650
150 kg N/hectar		3710	1710

This is mainly subsoil of heavy clay with very low content of
humus. Here it is difficult to obtain satisfactory yields.

The applications of sludge were very high and the response of
yield good, mainly due to better physical properties.

Measurements three years after sludge applications showed significant differences in the water-stability of soil aggregate
fraction 2-0,6 mm. The results for experiment 2 were as follows:

	Water-stability pst
0 tons sludge d.m./hectar	5
50 tons sludge d.m./hectar	13
100 tons sludge d.m./hectar	18

Sludge application to this soil has improved the ability to
resist hard rain and erosion.

2.3. Calculations on farmers'use of sludge

In relation to our investigations we have tried to calculate
the costs as to utilization of sludge for farmers. Thus, we provided that the dewatered sludge would be transported to the
farms without costs and stored there until the spreading of it.

			Kr/tons dry matter
Income		Nitrogen	35 N.Kroner
		Phosphorus	50 N.Kroner
		Potassium	÷ 10 N.Kroner
		Soil conditioner	0-50 N.Kroner
		Total value	75-125 N.Kroner
Expenses	Storage		
	Loading		
	Spreading etc.		
	Total expenses	Disadvantage 80 N.Kroner	

However, we have to make allowance for such accounts.

It is, anyhow, indicated that the fertilizer effect is at about
the same as the total expenses and disadvantages. The main
nettoincome is, therefore, due to the effect on physical properties.

We recommend that sludge is used on soil low in organic matter.
There are soils in all parts of the country which are in need
of better quality. The humus content of norwegian soil is,
however, on an average, high.

2.4. Farmers interest

Farmers are very interested in getting sludge. They consider
it a soil conditioner and they want to use as big quantities as
possible on poor soils.

2.5. Field lysimeters

Dos. Uhlen (1977) has carried out field lysimeter experiments on sloping cultivated land. The amount of surface runoff and erosion as dry soil in the period 13.7 - 8.9.1974 is shown in the table :

SURFACE RUNOFF AND SOIL LOSSES IN SUMMER TIME

	Surface runoff mm	Dry soil tons/ha
Exp. I 9 per cent slope		
Arable soil, without sludge, grain crop	56	0,9
" " with " " "	15	0,2
Fallow plot	58	6,3
Exp. II 4,5 per cent slope		
Arable soil, without sludge, grain crop	30	0,2
" " with " " "	12	0,1
Fallow plot	71	4,0

The applications of sludge were yearly 15 tons dry matter per hectar, in experiment I applied 3 years and 2 years in experiment II.

Applications of sludge have increased the <u>infiltration capacity</u> of the soil. This has <u>reduced the surface run-off</u> and given a <u>better resistance against soil erosion</u>.

3. PHYSICAL PROPERTIES IN SLUDGE AND SLUDGE COMPOST

For a good growth medium the following is needed :
- it should be able to store water and nutrients;

- it should provide sufficient amounts of air;

- excess of water has to percolate;

- it should be easy to cultivate;

- it should be a stable ground for plants;

- it should resist erosion.

The organic material in soil will in most cases contribute to good physical properties. Yet, there are examples that this is not the case. Fuel peat for instance, is far more impervious than desirable, while pure bark has not got sufficient ability to store water.

Fresh sewage sludge is extremely impervious. Coarse particles, gravel and sand are removed by initial treatment of sewage water. The inorganic matter, which in digested sludge represents about 60 weight percent or more, consists of silt and clay fractions. The organic matter in fresh sludge consists of fine-grained particles. It is a question of judging, whether sludge is too impervious as a growth medium, and thus best fitted as soil improvement on coarse soil types, but not on clay soils.

To elucidate that question several investigations have been made as to physical conditions in sludge.

3.1. Structure

Fresh sludge has no visible structure of aggregation. We have a so-called single-grain structure. The mass is pasty in natural dewatered state. It resembles fuel peat.

Quickly dried fresh sludge is disposed to shrink much in the upper layer. A firm and very hard crust is likely to appear.

Fresh sludge is not well apted for geen areas. It is difficult to level, and machines might easily spin and get stuck in the pasty mass. Fairly moist material laid out in summer might cause firm clods.

Composting of sludge will create an aggregation of single particles. The mass will therefore consist of grain and small clods.

An investigation of 4 types of compost shows that the fraction 6-20 mm represented 40-60 % of the mass, while there were 2-15 % of coarser fractions. The coarser particles represent a bigger part of it than you would wish. It is possible that crushing clods could be needed in cases where pure sludge is composted. Probably, most clods were formed by composting of slightly dewatered sludge. The quantity of clods is supposed to be reduced by further decomposing and treatment.

Experiments have also been made to find out about the aggregate stability by means of artificial precipitation. These are showing that the aggregates are far more stable than for instance aggregates in clay soils. Even though a fractioning of clods might occur, aggregat structur in sludge compost is hardly destructed.

3.2. Hydraulic conductivity

Hydraulic conductivity has been measured in several types of sludge and compost, which has reference to saturated flow both at natural storage and storage in laboratories. Since the measurements have shown a fairly large variation, only the main results are given here :

	Hydr. conducticity cm/hr
Raw sludge	0.01-0.04
Sludge compost	20-150
Raw sludge, stored 5 mth , bottom layer	0.01
Raw sludge, stored 5 mth, upper layer	20
Anaerobic sludge, fresh	0.01
Anaerobic sludge, stored 4 mths, upper layer	10
Anaerobic sludge, stored 3 years, 50 cm depth	0.5
Anaerobic sludge, stored 3 years, 30 cm depth	20
Anaerobic sludge, stored 3 years, upper layer	26

Consequently hydraulic conductivity is several thousand times higher in well decomposed sludge than in fresh sludge. To make comparisons we can mention that sandy soil has a hydraulic conductivity of 0.6-60 cm/hour.

The classification system of soil permeability rates is given in table 1 (KOLNKE 1979).

TABLE 1 : CLASSIFICATION OF SOIL PERMEABILITY RATE (KOLNKE, 1979)

| Descriptive term | Percolation rate | |
	in/hr	mm/hr
Rapid	> 6.3	> 160
Moderately rapid	2.0 - 6.3	50 - 160
Moderate	0.63 - 2.0	16 - 50
Moderately slow	0.20 - 0.63	5.0 - 16
Slow	0.05 - 0.20	1.25- 5.0
Very slow	< 0.05	< 1.25

Decomposed sludge appears to be very permeable to water. It has a great infiltration capacity. On the other hand fresh sludge is practically impermeable. Even at a low precipitation intensity the major part of the water will run off on the surface. Evidently, a radical change in the structure has taken place by the composting process.

This is illustrated in figure 1.

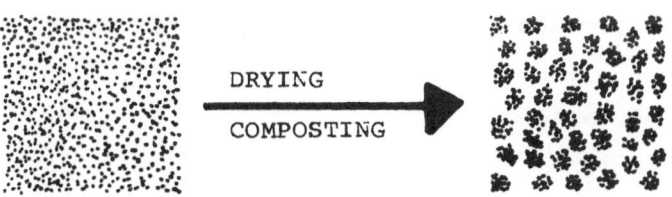

Singel particels   Aggregates "grainstructure"

Figure 1 : Changing in aggregate structure by composting

The change in hydraulic conductivity in bottom and upper layers
of stored sludge are mainly due to different rates of decomposition. To some extent conductivity may become reduced because of
pressure from over-lying masses.

There have been made experiments on sludge as dayly covering
layer on sanitary landfills. Driving on this material by continous supply of new garbage brought trucks into spinning in
rainy weather. Consequently, it is evident that high permeability
is considerably reduced by heavy pressure. Under such circon-
stances the structure is not firm enough.

We have also made simple measurements on infiltration velocity
in fresh sludge, frozen and unfrozen. The samples were not
water-saturated and the results were :

Fresh raw sludge, unfrozen	0.01 cm/hr
Fresh raw sludge, frozen	100 cm/hr

The figures tell that freezing of sludge creates a completely
different structure. On the other hand we have experienced that
some of this open structure is easily destructed by mechanic
treatment.

Frost disintegration leads to a better aeration. This promotes
a biological digestion, which results in a more stable aggregate
structure.

3.3. Water-retaining capacity

Field capacity measurements have been made for sludge - and soil
types. Results are given in percent water of dry matter :

sandy soil (medium sand)	30
clay soil	50
fresh raw sludge	290
fresh lime-sludge	280
anaerobic sludge, stored 6 mth	250
anaerobic sludge, stored 3 years	210
sludge compost	210

The fresh sludge is probably represented by slightly too high values on account of minimal hydraulic conductivity and consequently retarded drainage.

It is obvious that sludge has <u>large ability to store water</u>. The results indicate that sludge compost can retain 4 times as much water as clay soil. This ability is primarily associated with the organic matter. By supplies of sludge the ability of mineral soils to store water will increase.

3.4. <u>Water availability</u>

pF-measurements have been used to elucidate how strongly water is bound in sludge samples. Water availability is listed below as the difference between wilting point (pF 4.2) and field capacity (pF 2.0). The results are compared with average values for normal soil types :

	Water availability vol %
fresh anaerobic sludge	39
composted sludge	49
compost, slightly mixed with sand	46
sand	6
sand containing org. matter	12
clay soil	20
peat soil	23

The measurements indicate that a large part of the water is available to plants, which part is much bigger than the one entering other types of soil. In slightly decomposed sludge water seems to be very strongly bound, i.e. to a larger extent than in composted sludge.

Sludge has evidently good capacity as well as to storing of water as to supplying of plants.

3.5. Aeration, field trial

The sowing of different sorts of plants in pure, anaerobic
sludge, stored during winter, caused certain growth damages.
This sludge had a very propitious structure due to ground frost.
After soil treatment it seemed comparatively fine-grained. We
concluded, therefore, that growth damages resulted from in-
sufficient aeration to plant roots.

In order to make sure about our presumption a field trial was
laid out. The following treatments were compared.

a) crumbling all clods, seed loose covered with soil
b) like a) but firm packing
c) hacking to provide as much clods as possible after sowing
d) like c) but before sowing.

The aeration achieved in the root zones of the plants was very
unequal. The trial showed the most uniform germination and the
best plant growth on replication b) where we figured the
poorest aeration. Consequently, it looks as if fairly fresh,
frozen sludge is able to provide sufficient aeration. Prospec-
tive growth damages are probably due to other factors, e.g. in-
sufficient decomposition, ammonia toxicity etc.

3.6. Shrinkage of frozen and unfrozen sludge after drying

Fresh sludge, which is quickly dried, becomes a very firm crust,
most unfavourable to sludge use.

In order to elucidate frost effect on drying progress, we have
on a small scale measured shrinkage of frozen and unfrozen raw
sludge after drying at different temperatures.

Sludge structure changed character totally by freezing and
thawing. While the unfrozen sludge had a slippery surface almost
impermeable to water, the frozen had become more filt (blotting
paper-) like and extremely permeable.

When dried partly, the frozen sludge got a lighter colour, and grew water-repellent. However, it was easily made moist, and the infiltration capacity was soon as good as before. The difficulty to become moist increased with the length of the drying process.

The surface of the unfrozen sludge became continously firmer as the drying proceeded, and by total drying it was firm as con- crete. The frozen sludge then had a structure comparable with a rusk.

The frozen sludge seemed to keep some of the open structure due to frose disintegration, and the shrinkage was here considerably reduced. This is demonstrated in figure 2.

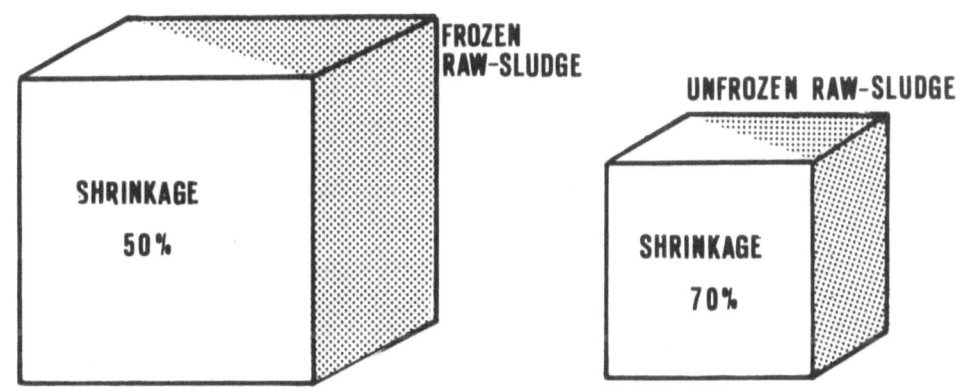

Figure 2 - Shrinking of raw-sludge after drying

4. LITERATURE

(1) KOLMKE, H. 1979. Soil physics. Tata McGraw-Hill publ. comp. LTD, New Delhi.

(2) NJØS, A. 1978. Behov for jordforbedring. Høgskoledagene ved NLH, stensiltrykk.

(3) UHLEN, G. 1977. Nutrient leaching and surface runoff in field lysimetres on a cultivated soil. Agr. Uni. of Norway. Dep. of Soil Fertility and Manag. Report No. 96.

DISCUSSION

Dr CATROUX :   Can you explain with more details the calcula-
                 tion of the value of sludge as soil condi-
                 tionner ?

Dr VIGERUST :   It is difficult to find the effect of sludge on
                 soil properties isolated from the fertilizer
                 effect. Based on all our experiments, we sug-
                 gest the value of sludge as soil amendment will
                 vary from 0 for soils with high content of or-
                 ganic matter - up to 50 N kroner per ton dry
                 matter for soils with bad physical properties.

Dr DE HAAN :   How can the K effect be negative ? Sewage sludge
                 does not contain much K, but always some; so
                 there should be a positive effect ?

Dr VIGERUST :   When farmers shall use sludge there must be a
                 plan for fertilizer for 3-5 years. In this way
                 it will cause a little practical problem to the
                 farmers when heavy applications of N and P. We
                 have ideally combined fertilizer high in K and
                 low in N and P.

Dr DIEZ :   You compare benefits and expenses of sludge use.
                 Do you charge the farmer or the sewage plant
                 with the expenses ?

Dr VIGERUST :   There has to be an agreement between the sewage
                 plants and the farmers about the expenses of
                 sludge distribution. Until now we have
                 recommended that sludge should be brought to
                 the farmers who have to store it, load it and
                 spread it. We think this will work.

Dr DIEZ :   Don't you have any problems with heavy metals,
                 because you recommend such high amounts of
                 sewage sludge.

Dr VIGERUST :   We allow to apply relative high amount of sludge because :

1. The heavy metal content of our sludges is relatively low.
2. Calculations shows that the heavy metal problem in our society seems to be low.
3. The sludge is mostly used to soils where it is onesided cereal-production. The grain will be mixed from many areas.

    This will give a very effective dilution.

Dr MOREL :   Do you spread sludges in winter as Canadians ?

Dr VIGERUST :   We can transport the sludge to the farms during winter where it can be stored in a safety way. Due to risks for water pollutions it will not be allowed to spread it.

Dr WILLIAMS :   Are you suggesting that sludge should only be applied to very low organic matter soils as there might be only very sandy soils and yet there could be clayey textured soils with structural problems which would benefit from sewage sludge ?

Dr VIGERUST :   Many of our soils which had bad physical properties are clay and loam. But also sandy soil can be improved by applying sludge. These soils are however more sensitive for the heavy metal - problem than clay soils.

Dr BERGLUND :   A comment.
Not only sandy soils will benefit from organic matter in sludge but even other soils with low range of organic matter such as morain soils and heavy clay.

MODIFICATIONS OF SOME PHYSICAL AND CHEMICAL SOIL

PROPERTIES FOLLOWING SLUDGE AND COMPOST APPLICATIONS (*)

G. GUIDI, M. PAGLIAI AND M. GIACHETTI

1. INTRODUCTION

To maintain a good level of fertility the soil needs periodical additions of organic materials. Sludge contains about 50 % organic matter and therefore can be used in agriculture wherever, as for instance in Italy, there is a shortage of manure and the climate causes a quick decomposition of any organic material added to soil. Changes in physical condition and chemical composition of the soil may be induced by sludge addition and should also be considered in an agricultural utilization of such materials.

The objective of this study was to determine the effects of aerobic and anaerobic sludges and their composted mixtures with the organic fraction of urban refuse on some physical and chemical soil properties under natural field conditions. Modifications occurring in total porosity, water stability index and electrical conductivity were investigated.

2. MATERIALS AND METHODS

A field study was established in May 1978 on a sandy loam soil which contained 0.9 % organic matter, 10.0 % clay, 14.1 % silt

(*) This work was supported by the National Research Council, Special Research Project "Environmental Quality Promotion".

and 75.9 % sand. The pH in water was 5.8 and C.E.C. (meq/100 g) 13.4. The experiment was carried out on 500 m^2 plots planted to corn (Zea mays L.).

The treatments included aerobic sludge (AS), anaerobic sludge (ANS), compost of aerobic sludge and the organic fraction of urban refuse (40-60%) (CAS), compost of anaerobic sludge and the organic fraction of urban refuse (20-80%) (CANS) and manure (M). There was also a control plot (C).

Organic materials were surface applied before the seedbed preparation and ploughed in. The addition rates were calculated on the organic carbon basis and were equivalent to 50 and 150 tons/ha of manure. Soil samples were taken periodically from 1978 to 1980.

<u>Total porosity</u> - Undisturbed soil samples (six for each plot) were air dried at room temperature, impregnated with a polyester resin and made into large (6 x 6 cm) thin sections (Jongerius and Heintzerberger, 1975). Thin sections were photographed (Pagliai et al., 1980) and each photograph analized by an electro-optical apparatus (Leitz Classimat).

<u>Water stability index of soil aggregates (WSI)</u> - A modification of the wet sieving method of Malquori and Cecconi (1962) was used to determine the stability of soil aggregates in water, since this method was sensitive enough to show up changes in soil aggregation (Giovannini and Sequi, 1976; Pagliai et al. 1978). Composite soil samples, made up of five subsamples, were taken to 20 cm depth from each plot, air dried and sevied out to obtain 1-2 mm aggregates. Three grams of soil aggregates were placed in a 0.25 mm mesh sieve and then moistened by the water rising by capillarity from a layer of wet sand. Wet sieving was carried out in deionised water with an alternative rotating movemnt (60 times per minute). The WSI is defined as 100(1-A/B), where A and B are the weights of aggregates passing through the sieve after 5 and 60 minutes, respectively.

Electrical conductivity - Electrical conductivity was determined
on the saturation extracts of soil samples following the usual
method (Black, 1965) soil samples were taken according to the
procedure just described for WSI, except that they were taken
at two depths : 0-20 cm and 20-40 cm.

3. RESULTS AND DISCUSSION

Mean values of total porosity, expressed as per cent of the
total area of each thin section occupied by pores are reported
in fig. 1. Total porosity increased significantly in all soil
samples taken in plots treated with any kind of organic materials,
irrespective of the sampling date. Generally, no significant
difference was observed in soil samples taken in plots treated
with the two different rates of organic materials. This means
that the rise in porosity was, for a given amendment, independent
of the rate of the amendment addition. While in 1978 total porosity increased over the period July-September, in 1979 the
trend changed and total porosity generally decreased over the
longer period June-November. This may be explained as the result of physical stresses caused by heavy autumn rainfall which
increased soil compactness. Samples were in fact taken before
the harvest to avoid any modification of soil compactness
brought about by harvesting activity. Increases of total porosity observed in 1978 may have been related to seasonal variations of bulk density which were attributed to biological
activity (Kladivko and Nelson, 1979).

WSI's of soil samples taken in plots treated with low and high
rate of organic materials are reported in fig. 2 and 3
respectively. In a soil quite rich in sand and poor in organic
matter, such as the experimental soil, the poor stability of
aggregates to water cannot be expected to be improved very much
by any organic material.In such a situation the increased granu-
lation and porosity of soil crumbs created by the organic matter
can easily be destroyed by water stresses because it is very

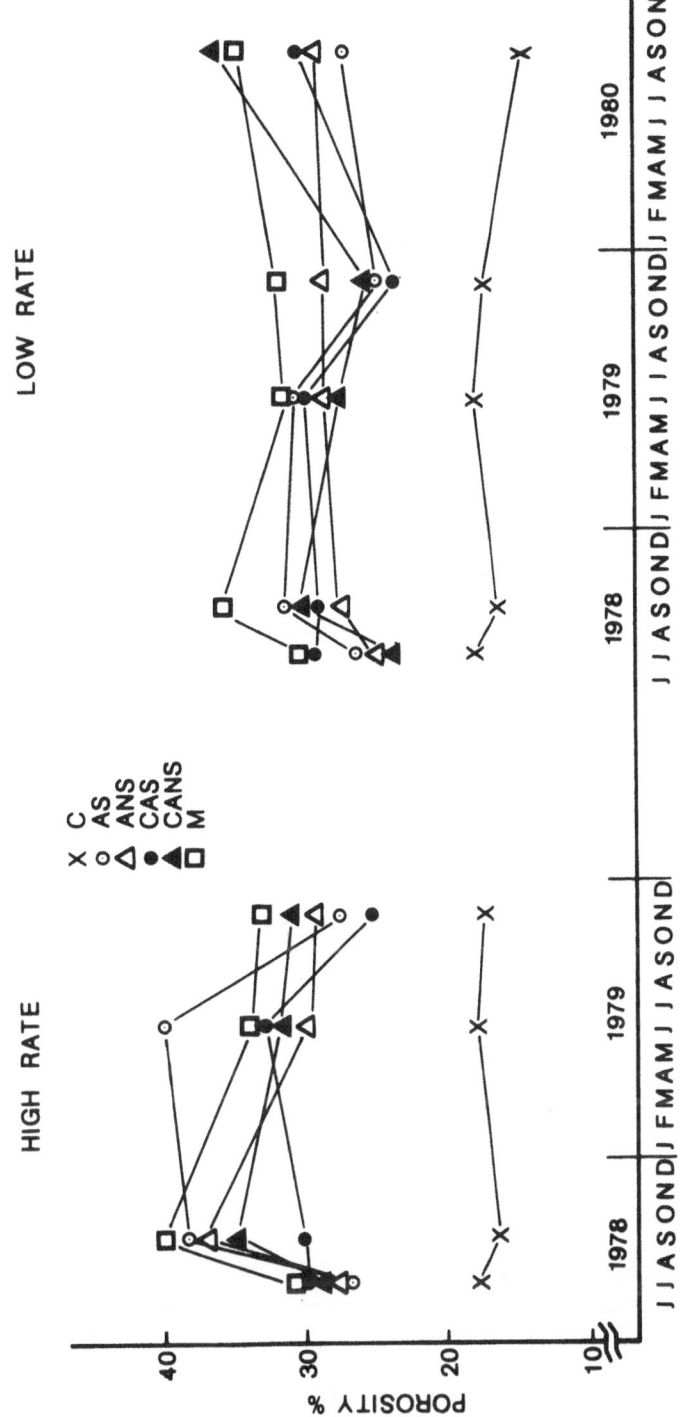

Fig. 1 : Effect of treatments on total soil porosity. Addition rates correspond to 50 (low) and 150 (high) tons per hectare of manure on the organic carbon basis.

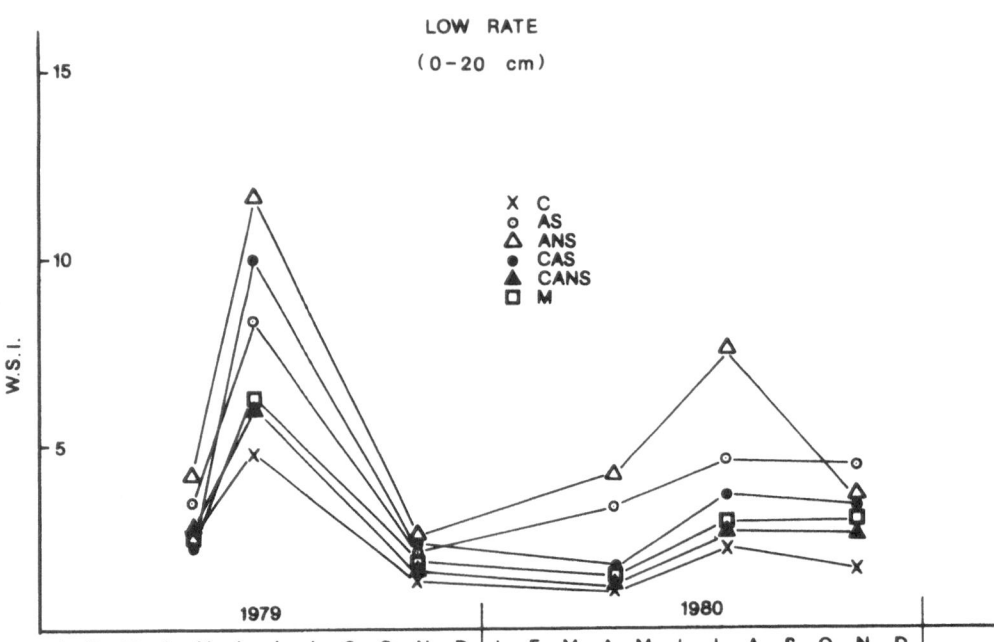

Fig. 2 : Effect of treatments on water stability index (WSI). Addition rate corresponds to
50 tons/ha of manure on the organic carbon basis.

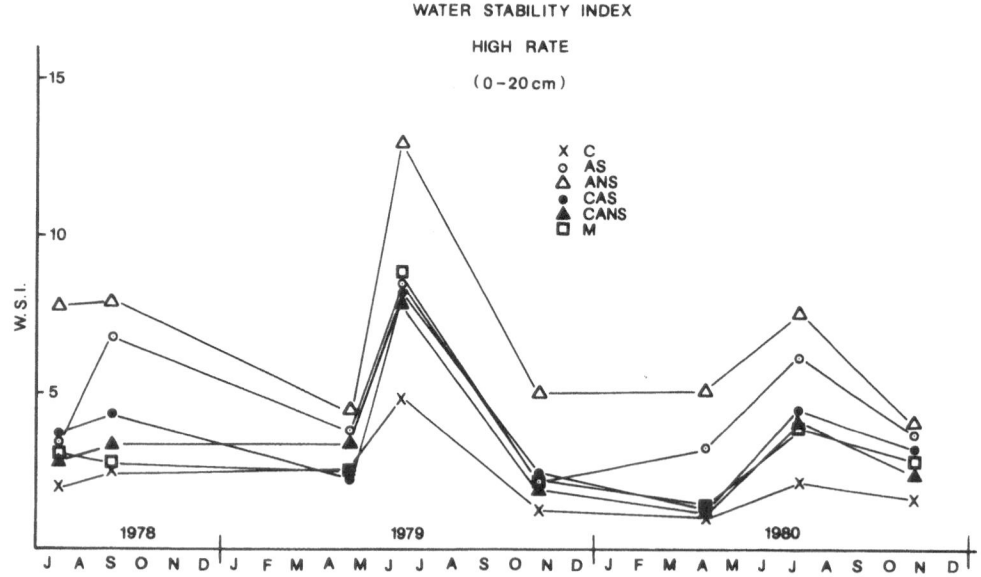

Fig. 3 : Effect of treatments on water stability index (WSI). Addition rate corresponds to
150 tons/ha of manure on the organic carbon basis.

difficult for organic compounds to succeed in binding together large particles of sand. Even though there were some differences in the extent of annual variations of WSI some common features were noticed. The WSI increased during the spring and early summer in correspondance with high biological activity existing in soil during this period. However the aggregating ability of sludges and composts decreased sharply after some months at the beginning of autumn rainfaill. As was previously observed for the total porosity, only in some cases the effect of the two different addition rates was evident on WSI.

An excess of soluble salts in the soil solution adversely effects germination and growth of plants. For this reason the EC was measured, at two different depths (0-20 and 20-40 cm), in those plots treated with higher rates of sludges, composts and manure. Values are reported in fig. 4 and 5. Soil salinity increased after all treatments but only aerobic sludge raised the value of EC to about 4 in the top soil. Figures for all other treatments were considerably lower, especially in samples taken at 20-40 cm depth. At the end of the growing season EC values dropped considerably as a result of laching and by the next season they were at levels tolerable for most plants.

In conclusion it can be said that sludges and composts improved soil porosity and water stability index of soil aggregates in a similar way to manure. Moreover, it is apparent that such improvements were practically of the same order for the two addition rates and this makes the use of the higer rate unnecessary.

The Authors wish to thank Mr. G. Lucamante and Mr. M. La Marca for their technical assistance.

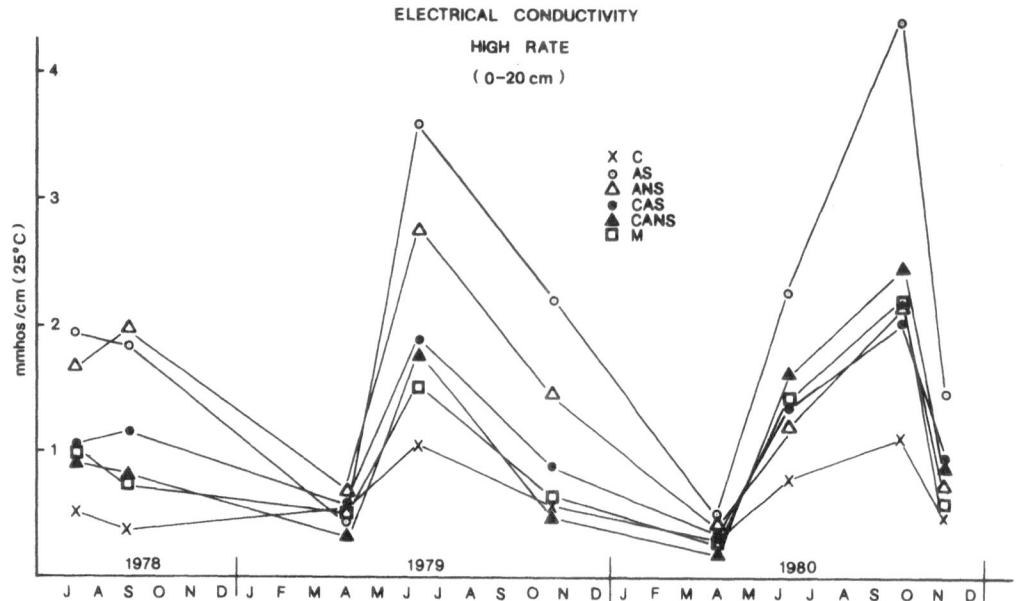

Figure 4 : Effect of treatments on electrical conductivity (EC)
at the O to 20 cm depth. Addition rate corresponds
to 150 tons/ha of manure on the organic carbon basis

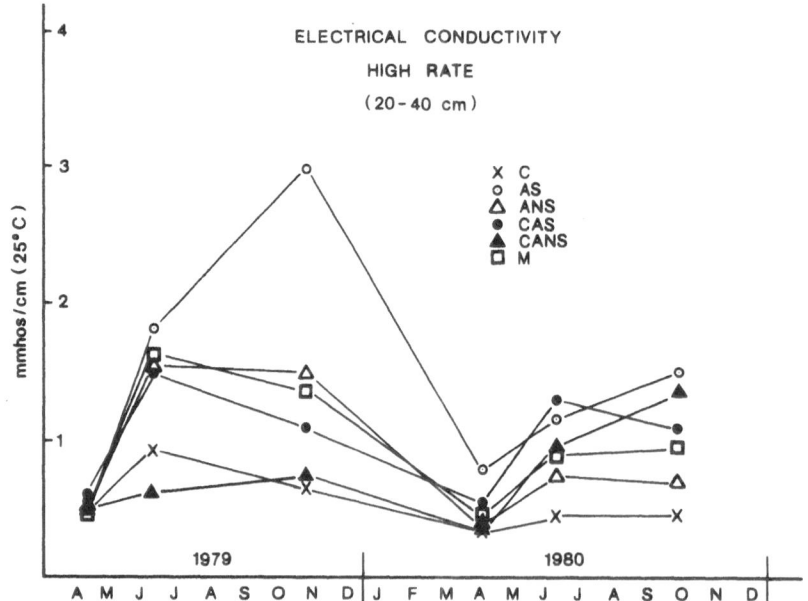

Figure 5 : Effect of treatments on electrical conductivity (EC)
at the 20 to 40 cm depth. Addition rate corresponds
to 150 tons/ha of manure on the organic carbon basis

4. ABSTRACT

Total soil porosity, water stability index of soil aggregates
and electrical conductivity were studied on a sandy loam soil
treated with sludges and composts since 1978. Addition rates
were equivalent to 50 and 150 tons/ha of manure on the organic
carbon basis.

Total porosity of treated samples was higher than the control
at all times and the overall effect was similar to that exerted
by the manure. Water stability indexes showed seasonal variations
in all plots. However soil aggregates of treated plots were
always more stable than those taken in the control plot.
Electrical conductivity increased in all plots following treatments but, in every case, winter rainfall lowered the content
of soluble salts to a level close to that of the control.

5. REFERENCES

(1) BLACK, C.A. 1965. Methods of Soil Analysis. American Society
    of Agronomy, Inc., Madison, Wisconsin, U.S.A.

(2) GIOVANNINI, G. and SEQUI, P. 1976. Iron and aluminium as
    cementing substances of soil aggregates. II. Change in sta-
    bility of soil aggregates following extaction of iron and
    aluminium by acetylacetone in a non-polar solvent. Journal
    of Soil Science 27 : 148-153.

(3) JONGERIUS, A. and HEINTZERBERGER, G. 1975. Methods in soil
    micromorphology. A technique for the preparation of large
    thin sections. Soil Survey paper N° 10, Netherland Soil
    Survey Institute, Wageningen.

(4) KLADIVKO, E.J. and NELSON, D.W. 1979. Changes in soil pro-
    perties from application of anaerobic sludge. J. Water Poll.
    Control Fed., 51   : 325-332.

(5) PAGLIAI, M., GUIDI, G. and LA MARCA, M. 1980. Macro and micromorphometric investigation on soil-dextran interactions. Journal of Soil Science, 31 : 493-504.

(6) PAGLIAI, M., GUIDI, G. and PETRUZZELLI, G. 1978. Effect of molecular weight on dextran soil interactions. p. 175-180. In W.W. Emerson, R.D. Bond and A.R. Dexter (ed.) Modification of Soil Structure. John Wiley and Co., New York.

DISCUSSION

Dr CATROUX :   Do you agree with the idea expressed this morning by Dr DANNEBERG and Dr DIEZ : the sludge can't be used practically for physical soil improvement ... and what occurs with nitrogen ?

Dr GUIDI :   No, I don't agree because the organic matter content of Italian soil is around 1 % and such soils need organic matter to maintain a good level of fertility. The low rate we used, equivalent to 50 tons/ha of manure, is quite normal as addition rate of organic matter in Italy and does not contain much more N than that required by crops.

Dr DANNEBERG :   A comment. The value of sludge for the farmer should be a higher yield or a save of fertilizers.

Dr DIEZ :   A comment to Dr CATROUXs' remark. In future in our country the dosis of sludge which is allowed to applicate will be restricted to 2.5 t dry solids per ha. Such small amounts will have only a fertilizing effect. The effect of the organic matter will disappear.

Dr FURRER :   How much P and N you applied per ha and per year with the high dose of sewage sludge ?

Dr GUIDI :   Roughly 400 kilos of total Nitrogen per ha and per year and the same amount of phosphorus (as P_2O_5).

Dr FURRER :   You found a total porosity of less than 20 % in the control. In our country, we find values lower than 43 % very seldom and we think, they are extremely bad. What the method, you determine the total porosity ?

Dr GUIDI :   The low values found are due to the particular
                 method used in this case which analyses porosity
                 in thin sections of soil.

Dr WILLIAMS :   Have you considered the effects of ionic concen-
                 trations up to 3 and 4 milliohms . cm^{-1} on aggre-
                 gation in the short-term ?

Dr GUIDI :   The S.A.R. (sodium adsorption ratio) in the sa-
                 turation extracts was always very low, and there-
                 fore the sodium ion present in the soil probably
                 did not have any detrimental effect on the
                 aggregation.

Dr CATROUX :   Have you any indication on the sludge content in
                 mineralizable carbon ? Have you some explanation
                 about the best stabilization effect with anaero-
                 bic sludge treatment ?

Dr GUIDI :   I am sorry, I haven't datas on mineralizable
                 carbon. The better stabilizing effect of anaero-
                 bic sludge could be attributed to the relatively
                 large amounts of stable organic compounds present
                 in anaerobic sludge, such as lignin, cellulose
                 etc. These compounds can either improve directly
                 soil structure by interactions with soil surfaces
                 or indirectly because they are rather resistant
                 to the microorganisms which decompose organic
                 materials in soils.

Dr DANNEBERG :   Do you agree with the interpretation that the
                 high stability produced by anaerobically digested
                 sludge is due to bigger microbial activity ?

Dr GUIDI: Yes, I agree.

Dr DIEZ :   The general opinion is that anaerobic sludge
                 affects microbiological activity negatively.

Dr DE HAAN :   Dr FURRER found a higher content of O.M. in aero-
                 bically digested sludges compared to anaerobically

digested sludges. Does that mean a greater decomposability of aerobically digested sludges (or a greater stability of anaerobically digested sludges ?)

Dr FURRER :   We think that the organic matter of aerobic digested sludges with a high organic matter content is easier decomposable. Also the organic nitrogen of these sludges is easier mobilizable as found in pot experiments.

Dr DIEZ :   Why do anaerobic treated sludges have a higher organic matter content than the anaerobic ones ?

Dr. FURRER :   In an average, the anaerobic digested sludges contain 41 % organic matter, the aerobic one 57 %. The anaerobic digestion takes place by $\simeq$ 35°C, produce energy (gas) and is done therefore as long as possible. On the other hand, aeration is energy-consuming and expensive. So it is done as short as possible.

Dr CHAUSSOD : I compared aerobic sludges and anaerobic sludges in laboratory conditions. And I think that the differences in nitrogen and carbon mineralization are not due to the way of stabilization, but much more to the degree of stabilization. There are aerobic sludges more or less stabilized and anaerobic sludges more or less stabilized.

EFFECTS OF SEWAGE SLUDGES ON SOIL BIOLOGICAL PROPERTIES

Effect of sewage sludge on soil humus content

Influence of sewage sludge application on organic matter content, micro-organisms and microbial activities of a sandy loam soil

Influence of sewage sludge application on microarthropods (collembola and mites) and nematodes in a sandy loam soil

Soil microbial activities as affected by application of composted sewage sludge

Side effects of sewage sludges : possible enhancement of denitrification

Land application of sludge : effects on soil microflora

Sludge effect on soil and rhizosphere biological activities

Some biological properties in sewage sludge and sewage treated soils

EFFECT OF SEWAGE SLUDGE ON SOIL HUMUS CONTENT

I. KOSKELA

1. INTRODUCTION

The addition of organic matter to sandy soils has improved soil
properties. At Agricultural Research Centre, Finland, we would
like to find out what happens, when sewage sludge is added to
soil with good growing condition.

Field experiment was started in autumn 1973. Amounts of sewage
sludge were 50 and 100 tons dry matter per hectare and were
broadcasted before ploughing. Digested sewage sludge from the
city of Helsinki was precipitated by $FeSO_4$, dry matter content
about 20 % and humus content about 50 % in dry matter. Type of
soil was sandy clay, pH 5-6. In addition to sewage sludge
(given only 1973) 50-100 kg N per hectare in NPK-fertilizer
(20:4;4:8.3) has been sown in drills every spring.

Every year it has been four different blocks (I-IV) :

Year	I	II	III	IV
1974	rye	sugar beet	turnip rape	barley
1975	barley	"	"	"
1976	hay	"	"	"
1977	"	"	"	"
1978	"	"	"	"
1979	"	"	"	"
1980	barley	barley	barley	"
1981	barley	"	"	"

Repeats : 4

2. YIELDS

Sewage sludge increased most of all barley yields. During the years 1974-1980 barley yields (block IV) were : (kg/ha)

		Sludge applied tds/ha	
Year	0	50	100
1974	4670	4500	4470
1975	2980	4490	4510
1976	4110	5300	5730
1977	3320	4390	4730
1978	2450	3550	3890
1979	3110	4190	4480
1980	2510	3410	3660
1974-1980	23150	29830	31470
		+ 6680	+ 8320

3. INFLUENCE ON THE SOIL HUMUS CONTENT

In autumn 1980 soil samples were taken from the depth of 0-20 cm. C-contents were measured in spring 1981 with LECO[R] CR-12. The results show that the soil humus content was increased markedly with sewage sludge application (Fig. 1).

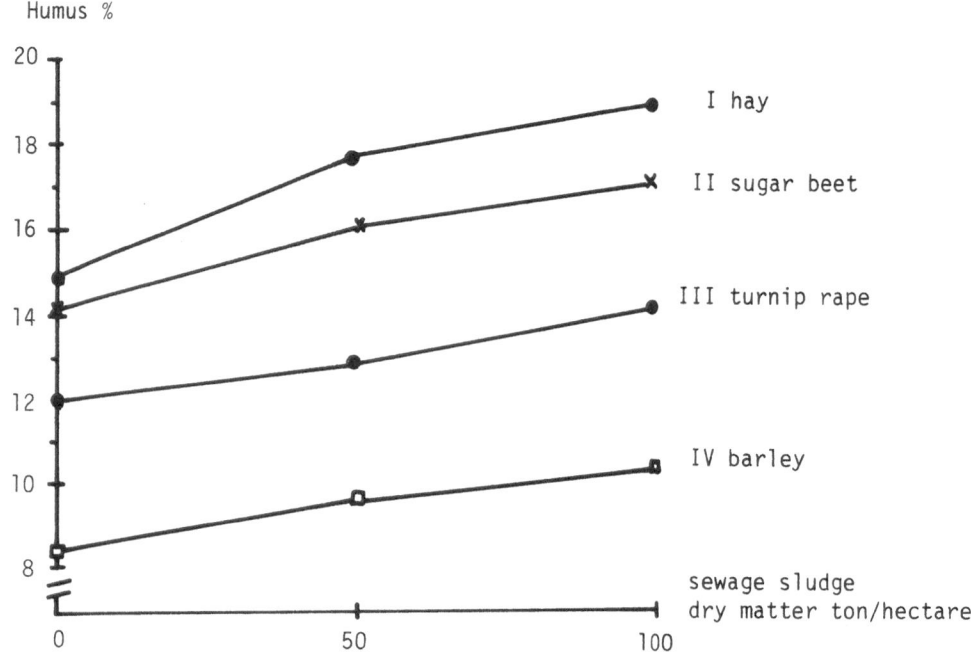

Figure 1 : The effect of solid digested sludge on soil humus
content (sandy clay). Humus content = C x 1.724

Differences between treatments and plants (blocks) were
significant.

It seems the influence of digested sewage sludge on the soil
humus content is significant even on the soil with much
organic matter and it last several years.

4. SUMMARY

Field experiment was started in autumn 1973 and it is still
going on. Amounts of sewage sludge were 50 and 100 ton dry
matter per hectare. Digested sewage sludge was precipitated by

$FeSO_4$, dry matter content about 20 %, humus content about 50 % in dry matter. Type of soil was sandy clay, pH 5-6. In addition to sewage sludge (given only in autumn 1973) 50 or 100 kg N per hectare in NPK-fertilizer has been sown in drills every spring.

After 7 years soil humus content were :

    a) no sewage sludge   12,40 %
    b) 50 tons sludge dry matter 14,09 %
    c) 100 tons sludge dry matter 15,14 %

DISCUSSION

Dr SÜSS :   Is your experiment running for long time ?

Dr KOSKELA :   It was started in autumn 1973.

Dr WILLIAMS :   Of the humus added in 100 t/ha of dry sludge
                 solids how much is remaining after 7 years ?
                 On a rough calculation from the organic matter
                 contents of the soil it would appear that a
                 hich percentage of the humus added is still
                 remaining.

Dr KOSKELA :   We have not calculated the amounts of humus re-
                 maining from the sludge application.

INFLUENCE OF SEWAGE SLUDGE APPLICATION ON ORGANIC MATTER

CONTENT, MICRO-ORGANISMS AND MICROBIAL ACTIVITIES

OF A SANDY LOAM SOIL

X. STADELMANN AND O. J. FURRER

1. INTRODUCTION

From ecological point of view, the disposal of sewage sludge
either by burning or incineration is not a satisfactory solution.
A high content of organic matter (approximately 43 % of the dry
matter) and a high content of plant nutrients (P, N, Ca and Mg)
speaks for its utilizaiton in agriculture rather than disposal.
At present, approximately 6 % of the total P requirement and
2.5 % of N fertilizer requirement (in form of manures and
fertilizers) of Swiss agriculture is covered by sewage sludge(10).
Prerequisite for the sewage sludge utilization is nevertheless
justifiable amounts of toxic substances (e.g. heavy metals) and
harmful organisms (e.g. Salmonellae).

A recirculation of sewage sludge in soils would be unthinkable
without microbial activities. In this case one thinks mainly
about the mineralization of sludge organic matter and about the
released plant nutrients such as CO_2, ammonia, nitrate, phosphate
and sulphate.

From the microbiological point of view, a land application of
sewage sludge as fertilizer places following set of problems :

1. How is the influence of sludge application on the humus content, humus quality and thereby related water, air and nutrient status in soils ?

2. What are the effects of sewage sludge fertilizer on the soil microbiological equilibrium (systematic and functional groups), the soil biological activities (enzyme activities, biochemical decomposition, synthesis processes), and thereby on the soil fertility, plant nutrition and plant dry matter yield ?

3. Does toxic substances (such as heavy metals) associated with the sewage sludge present a danger for the biochemical processes in soils ?

4. How big are the dangers, which are assessed on the basis of changes in biological activities (such as nitrate leaching, denitrification) to the environment due to the land application of sewage sludge ?

5. Are the soil microorganisms able to decompose the foreign substances (mineral oil, tar products, pesticides, detergents, oestrogens PCB and polycyclic carbohydrates) which could be added to the soil through sewage sludge application (18) ?

6. Is the soil able to inactivate the human and animal pathogens, which could get into it by land application of sewage sludge ?

7. What are the influences of sewage sludge application on the soil-born plant pathogens and parasites ?

The present paper mainly deals with the first three questions mentioned in above paragraph. The precise objective of this fields investigation was to clarify that how a fertile soil under many years of sewage sludge application at two different rates i.e. a normal and a heavy reacts with respect to :

- its humus content

- its microbiological population (bacteria, actinomycetes, yeasts and hyphal fungi) and soil microbiological equilibrium

- its soil microbiological activities (catalase) and the important mineralization processes (CO_2- evolution, ammonification, alkaline phosphatase).

2. MATERIALS AND METHODS

2.1. Experimental location, soil type and soil characteristics

In order to determine the humus content and the soil microbiological parameters, advantage of a field experiment (Bodenbe-
lastungsversuch mit Klärschlamm und Schweinegülle = sewage
sludge and pig slurry loaded soil) which was conducted in the
campus of Swiss Federal Research Station, Liebefeld-Bern was
taken. This station lies at 565 m above sea level with a average
annual rainfall and temperature of 1000 mm and 7.6°C respectively.
The experiment was laid out in April 1976 on a weakly developed
parabrown soil having good water holding capacity and is developed on the moraine gravel. Soil type (0-25 cm) : Sandy loam
(clay 16 %, silt 18 % and sand 66 %). Soil characteristics :
pH 5.7 - 7.0, total organic carbon 1.4 to 1.9 %, total Kjehldahl
N = 0.18-0.25 %, C/N ratio = 6.8 to 7.9, extractable
P = 3-17 µg/g, extractable K = 32-95 µg/g, $CaCO_3$ = 0-1.2 %,
cation exchange capacity (CEC) 12-20 meq/100 g.

2.2. Lay-out of the experiment

The experiment was laid out in such a manner that a normal dose
(2 t organic matter ha^{-1} y^{-1}) and a heavy dose (5 t organic
matter ha^{-1} y^{-1}) of liquid anaerobically stabilized sewage
sludge could be compared with mineral fertilizer and control
(unfertilized   treatment. The experiment was divided into   4
stripes. There was a crop rotation experiment on stripes A, B,
and C, whereas there was a permanent grassland experiment on
stripe D (table 1).

TABLE 1 : CROPS GROWN IN EACH STRIPE DURING EACH EXPERIMENTAL
          YEAR

Year	Stripes			
	A	B	C	D
1976	Silage maize	Winter wheat	Clover grass	Permanent grassland
1977	Winter wheat	Clover grass	Silage maize	Permanent grassland
1978	Clover grass	Silage maize	Winter wheat	Permanent grassland
1979	Silage maize	"Korn"	Clover grass	Permanent grassland
1980	Winter wheat	Clover grass	Silage maize	Permanent grassland

Each treatment has been repeated 4 times. Amounts of nutrients
which should be supplied by each fertilizer treatment are given
in table 2.

TABLE 2 : FERTILIZER TREATMENTS AND PROVISIONAL NUTRIENT DOSES
          SUPPLIED THROUGH THEM

Treatment	Amounts of nutrients supplied (kg ha^{-1}y^{-1})				
	O.M.	N	P	K	m^3/ha
Min = Mineral fertilizer	0	150-220*	63.6	249	-
0 = Without fertilizer	0	0	0	0	0
SS$_2$ = Sewage sludge normal dose	2000	(150)60**	69.8	14.6***	60- 80
SS$_5$ = Sewage sludge heavy dose	5000	(375)150**	174.6	41.5***	160-200

*   150 kg was supplied to silage maize and wheat/"Korn", whereas 220 kg
was supplied to clover grasse and permanent grassland.

**   The number presented in brackets are for total nitrogen (N_T), whereas
the values out side brackets are for available nitrogen (N_{av}) which is
calculated from the following formula :

$$N_{av} = 0.90 . N_{Am} + 0.25 . N_o$$

where N_{Am} = Ammonical nitrogen (NH_4-N)

      N_o = organically bound nitrogen

***   In case of sewage sludge treatments (SS$_2$ and SS$_5$) the values of supplied
K was not equal to the K supplied through mineral fertilizer treatment
(Min). Therefore K was compensated through mineral fertilizer.

Different amounts of sewage sludge were applied so as fixed amounts of organic matter could be supplied to the plots. And because of this fact there were yearly fluctuations in the amount of individual nutrients supplied to the plots. Table 3 shows the effective amounts of nutrients supplied to the stripe D (permanent grassland) through sewage sludge treatments.

TABLE 3 : AMOUNTS OF NUTRIENTS SUPPLIED IN THE STRIPE D (PER-
MANENT GRASSLAND) THROUGH SEWAGE SLUDGE TREATMENTS

Treatment	Year	Amounts of nutrients supplied (kg ha^{-1}y^{-1})				
		O.M.	N_T	N_{av}	P	K
SS_2	1976	2320	194	93	178	243*
	1977	1638	267*	189	71	149*
	1978	2005	236*	133	171	252*
	1979	2510	218	121	262	262*
	1980	1148	247*	185	133	259*
	$\bar{x}$	1924	232	144	163	253
SS_5	1976	4826	399	192	360	239*
	1977	4889	459	223	210	248*
	1978	4022	385	159	354	263*
	1979	6349	449	205	671	271*
	1980	4928	397	186	518	266*
	$\bar{x}$	5003	418	193	423	257

* These values include the amount of nutrient compensated through mineral fertilizer.

2.3. Soil sampling

In order to record changes in the humus status of plots receiving sewage sludge, soil samples (0-20 cm) were collected in spring 1976 (start of experiment), in autumn of 1977 and 1980. Mixed samples were prepared out of 4 replications per treatment. Organic carbon content was determined on mixed soil samples.

To record yearly changes in the soil microbiological parameters in the permanent grassland (stripe D), 19 soil samples were collected each year from each plot just before the vegetation period (sampling dates were : 25.7.1977, 13.3.1978, 19.3.1979, 10.3.1980 and 16.3.1981). Later on, a mixed sample was prepared out of these 19 samples and passed through a 2 mm sieve.

A comparison of soil microbiological parameters between arable land (stripe A) and permanent grassland (stripe D) soil samples which were collected on 27.10.1980, was done. In this case, soil samples (0-15 cm) from the treatment Min (3 plots/treatment/crop) and the heavy dose of sewage sludge (SS_5) were investigated. Soil sampling technique and preparation procedure were same as described in case of permanent grassland experiment.

2.4. Determination of organic matter and microbiological parameters in different soil samples

The organic carbon (C) content of the air dried soil was determined by dry combustion method (600°C) with the help of Carmograph-8 (Firm-Wösthoff, Bochum).

Detailed description of the soil microbiological procedures used for these samples is given by STADELMANN & FURRER (20). Field moist soil samples were used for the determination of micro-biological counts and CO_2 evolution. Whereas, air dried soil samples were used for catalase, alkaline phosphatase activities and ammonification. The quantitative determination of bacteria, actinomycetes, yeasts, hyphal fungi and algae were determined by the methods as described by STADELMANN (19). Catalase activity was measured by gas-volumetric determination (BECK (3)), whereas CO_2 evolution was determined by titration (JAEGGI (14)). The ammonification was determined by potentiometric method. The ammonification in soil samples which were collected during 1977-1979 was carried out in aerobic environment (4) whereas, from 1980 it was done under anaerobic condition (25). The phosphatase activity was determined by a modified procedure of VDLUFA.

2.5. Presentation of data and statistics

The humus content (organic C) is presented in percent and calculated on the basis of oven dry soil (tables 4 and 5). Microbiological data all together are based on 1 g oven dry soil.
The microbiological values are arithmetical mean with an exception of algal counts. In case of algae, the value are median.
If a treatment is significantly different than the treatment
Min in Wilcoxon-Range test (W-test p = 5 %), then the treatment
is marked with a star (*).

3. RESULTS AND DISCUSSION

3.1. Humus status of soil

The opinions are quite apart whether the sewage sludge directly
or indirectly could improve the humus status of soils and could
be used as soil conditioner (19). With a single application of
sewage sludge, one should not expect a significant improvement
in the humus balance of soil with respect to long term (17, 21,
23), although it would be different if heavy doses and repeated
application of sewage sludge are given (2, 6, 8, 9, 16). The
results of the field experiment described here have shown that
the changes in humus level are closely related not only to the
sewage sludge doses but also to the management practices and
also to the crop rotation. Following tendencies are recognizable
from the results presented in tables 4 and 5 :

- Humus status of <u>permanent grassland soil</u> has been improved
  through sewage sludge application. The improvement in humus
  content is related to the amount of applied sewage sludge
  which has already been shown by FURRER (9). There is an increase in the humus content in surface soil to the tune of
  2.08 t C/ha due to an input of 5.64 t C/ha applied in form
  of sewage sludge. In case of 14.21 t C/ha input (SS_5), there
  is an increase in carbon level of soil by 8.84 t C/ha (table 5).

Apparently, during a period of 4 1/2 years only 5.37 t C/ha
has been mineralized in case of heavy dose of sewage sludge
application. This is not a real interpretation. In this case
one should also consider the fact that the significant increase in C content (8.84 t C/ha) of permanent grassland
(with a nominal loss of 5.37 t C/ha) is not only due to direct input of organic C through sewage sludge but it is
partly also due to some of the indirect effects as well. The
fact that the nutrient locked in sewage sludge influenced the
dry matter yield of plants and thereby helped in increasing
the C content in form of rests of roots and litter. This idea
could be well substantiated from the results of mineral
fertilizer treatment (Min) (table 5).

- In unfertilized permanent grassland surface soil, during a
  period of 4 1/2 years, there was a loss of 6 t of C/ha from
  a total 42 t C/ha. This remarkable C loss from the soil could
  be due to reduced plant growth and a larger proportion of
  clover grass as compared to grass. Due to reason stated later,
  the surface soil received less amount of plant root rests as
  a source of organic substance. It is also likely that a portion of native humus was also consumptively decomposed by nitrogen fixing bacteria because nitrogen fixation is a extremely high energy demanding process (26) and for that purpose
  microoganisms have to attack native carbon source as well.

- In crop rotation experiment where partly the land was left
  fallow, the changes in humus status is also influenced by the
  amount of organic substances supplied through sewage sludge
  and also from the type of crops grown on it. After 4 1/2 years,
  there is significant increase in humus content due to sewage
  sludge application specially in case of crop rotation having
  cereals ("Korn"). The cereal crops either does not affect or
  slightly enhance the humus level of soil (7). In the crop
  rotation where silage maize was grown, there is no distinct
  increase in the humus content of soil. This means, that during
  4 1/2 years time, at least a minimum of organic C, added in
  form of sewage sludge (13.90 t C/ha), was mineralized
  (15.46 t/ha) (table 5). A slight consumptive decomposition of

TABLE 4 : CARBON CONTENT OF SURFACE SOILS (0-20 cm) AS INFLUENCED BY CROPPING PATTERN, FERTILIZER TREATMENT AND DURATION OF THE EXPERIMENT.

Stripes	Cropping pattern	Crops in 1977 and 1980	Fertilizer treatment	C-content (%)			Relative C-content (%)		
				1976[1]	1977[2]	1980[3]	1976[1]	1977[2]	1980[3]
A	Crop rotation (3 years)	Winter wheat/ "Korn"	Min	1.37	1.19	1.40	100	87	102
			0	1.31	1.31	1.32	100	100	101
			SS_2	1.21	1.19	1.44	100	99	119
			SS_5	1.35	1.26	1.56	100	93	115
B	Crop rotation (3 years)	Clover grass	Min	1.27	1.19	1.15	100	94	90
			0	1.21	1.19	1.17	100	99	97
			SS_2	1.27	1.21	1.20	100	96	95
			SS_5	1.16	1.24	1.39	100	107	120
C	Crop rotation (3 years)	Silage maize	Min	1.36	1.38	1.38	100	102	101
			0	1.37	1.35	1.40	100	99	102
			SS_2	1.50	1.46	1.44	100	97	96
			SS_5	1.55	1.45	1.49	100	93	96
D	Permanent grassland	Clover grass	Min	1.53	1.64	1.58	100	107	103
			0	1.63	1.58	1.40	100	96	86
			SS_2	1.53	1.59	1.61	100	103	105
			SS_5	1.57	1.66	1.91	100	105	121

[1] before start of the experiment — [2] 1 1/2 years after begin of the experiment — [3] 4 1/2 years after begin of the experiment

TABLE 5 : NATIVE AND ADDED SEWAGE SLUDGE ORGANIC MATTER (t C/ha) IN THE 0-20 cm SURFACE SOILS AS INFLUENCED BY THE CROPPING PATTERN AND FERTILIZER TREATMENT

Stripes	Cropping pattern	Crops in 1977 and 1980	Fertilizer treatment	1 Soil 1976	2 Addition of sewage sludge	3 1 + 2	4 Soil 1980	5 4 - 1	6 4 - (1 + 2)
A	Crop rotation (3 years)	Winter wheat/ "Korn"	Min	35.62	0	35.62	36.4	+ 0.78	+ 0.78
			0	34.06	0	34.06	34.32	+ 0.26	+ 0.26
			SS2	31.46	5.44	36.90	37.44	+ 5.98	+ 0.54
			SS5	35.10	13.30	48.40	40.56	+ 5.46	- 7.84
B	Crop rotation (3 years)	Clover grass	Min	33.02	0	33.02	29.9	- 3.12	- 3.12
			0	31.46	0	31.46	30.42	- 1.04	- 1.04
			SS2	33.02	5.60	38.62	31.2	- 1.82	- 7.42
			SS5	30.16	14.26	44.42	36.14	+ 5.98	- 8.28
C	Crop rotation (3 years)	Silage maize	Min	35.36	0	35.36	35.88	+ 0.52	+ 0.52
			0	35.62	0	35.62	36.4	+ 0.78	+ 0.78
			SS2	39	5.71	44.71	37.44	- 1.56	- 7.27
			SS5	40.3	13.90	54.20	38.74	- 1.56	- 15.46
D	Permanent grassland	Clover grass	Min	39.78	0	39.78	41.08	+ 1.3	+ 1.3
			0	42.38	0	42.38	36.4	- 5.98	- 5.98
			SS2	39.78	5.64	45.42	41.86	+ 2.08	- 3.56
			SS5	40.82	14.21	55.03	49.66	+ 8.84	- 5.37

1 Calculated native C-content in t/ha in the surface soil (density 1.3 kg/dm³) at the time of the start of the experiment (1976)

2 Calculated added C (in t/ha) through sewage sludge during 1976-1980 (loss of ignition: organic C=1.76:1 after GUPTA 1976 (13)

3 Expected amount of C which should have been present in soil in 1980 when there is no mineralization and no addition of C in form of roots and litter

4 Calculated C-content in t/ha in the surface soil 4½ years after the begin of the experiment

5 Net C-gain and net C-closs in t/ha in the surface soil after 4½ years of the experiment

6 Net C-mineralization loss and net C-mineralization gain (in t/ha) after 4½ years of the experiment

native humus due to the applied sewage sludge could not be excluded in case when the organic C supply through the sewage sludge and plant residues does not fully satisfy the demand of soil biomass for carbon and energy.

However, it would be possible to make clear statements about the influence of sewage sludge on the humus status in soil after lapse of few more years of experiment.

3.2. Soil microbiological investigations on permanent grassland

3.2.1. Heterotrophic and autotrophic soil microorganisms

In permanent grassland, sewage sludge application caused a distinct increase of the counts of saphrophytic and heterotrophic soil microorganisms, which are responsible for the mineralization of organic substances (table 6). The aerobic bacteria and actinomycetes profited from the sewage sludge application more than the yeasts and hyphal fungi with respect to their counts but not strongly with respects to their biomass. The heavy dose of sewage sludge (SS_5) increased distinctly the microbial populations as compared to normal dose (SS_5). The normal sludge dose (SS_2) increased the bacterial counts up to 59 %, the actinomycetes up to 78 %, the yeasts up to 106 % and the units of hyphal fungi up to 43 % as compared to mineral fertilizer treatment (Min). On the other hand, soil samples from the plots treated with heavy dose of sewage sludge (SS_5) approached up to 104 % more aerobic bacteria, up to 151 % more actinomycetes, up to 160 % more yeasts and up to 73 % more units of hyphal fungi in compare to mineral fertilizer. The increase in the counts of heterotrophic microoogranisms in sewage sludge treated plots was independant of experiment duration that means the number of years in which sewage sludge was applied continually and repeatedly as compared to mineral fertilizer treatment (Min).

TABLE 6 : INFLUENCE OF SEWAGE SLUDGE APPLICATION ON THE COUNTS
OF SOME IMPORTANT GROUPS OF MICROORGANISMS IN A PER-
MANENT GRASSLAND SOIL. AVERAGE OF 4 REPLICATIONS
TREATMENT^{-1} YEAR^{-1}. (Min = Mineral fertilizer
treatment = 100)

Year	Treatment	Aerobic bacteria	Actinomycetes	Yeasts	Hyphal fungi	Algae
1977	0	94	100	152	145*	338
	SS_2	128	140	180	123	19
	SS_5	190*	180	260	120	38*
1978	0	66	74	156	122	27
	SS_2	111	85	97	143*	57
	SS_5	183*	142	93	138	27
1979	0	87	53	170	120	100
	SS_2	128	77	150	113	46
	SS_5	204*	159	157	117	27*
1980	0	104	107	163	116	56*
	SS_2	123	160*	120	129	23*
	SS_5	174*	186*	147	134*	19*
1981	0	75	106	160	123	n.d.
	SS_2	159*	178	206	134*	n.d.
	SS_5	164*	251*	208*	173*	n.d.

*   = Significantly different than treatment - Min with W-test
0   = Without fertilizer

SS_2 = Sewage sludge normal dose
SS_5 = Sewage sludge high dose
n.d. = not yet determined.

The significant increase in the heterotrophic soil microoganisms
could be attributed on the one hand to the high content of
organic matter in sewage sludge which is a rich source for nutrients and energy and on the other hand to indirect fertilizer
effect (there is large amounts of roots and litter left over in
the field). It is worth mentioning here that in contrast to
several other reported experiments (11, 16, 19, 21, 22), in this

experiment not only the bacteria profited from sewage sludge application but also actinomycetes and fungi. From the point of view of soil microbiology, the increase in the heterotrophic soil microorganisms due to sewage sludge application could be judged favourably. Inspite of the increase in the fungal population in the plots treated with sewage sludge, there was a clear improvement in ratio of soil bacteria/soil fungi which is one of the soil fertility index (1). In control plots (unfertilized plots) there is an excessive increase in the fungi and yeast counts which has significantly reduced the ratio between bacterial counts/fungal counts -and thereby the soil fertility.

In contrast to the heterotrophic microorganisms, sewage sludge has drastically reduced the autotrophic microorganisms (table 6) in the surface soils (0-10 cm). A reduction of 1.7-5.2 times and 2.6-5.3 times of the green/blue green algae in case of normal and heavy doses of sewage sludge, respectively as compared to mineral fertilizer was observed. However, it is left to be clarified whether the sewage sludge application has startled primarily the green algae or also the nitrogen fixing blue algae. A reduction in the biological nitrogen fixation, which according to HANSON (13) is not expected to happen due to sewage sludge application, could be an ecological problem. It is still unanswered why the application of sewage sludge, reduced the soil algae in surface soil. This does not seem to be due to the heavy metals in the sewage sludge, which is substantiated by the fact that the control plots (unfertilized) have also low algal population. Nitrogen supply and soil microecological and microclimatic reasons could be considered in order to explain the phenomenon. The dark colour of the soil, which occured due to applied sewage sludge might have helped in the increase of soil temperature and also prevent the intruding sunlight into the surface soil layer, which are the important prerequirement for the normal algal growth.

Even i , in absolute terms, there is an intense increase in the heterotrophic soil microorganisms and an intense reduction of autotrophic soil algae due to application of sludge, it is still

to be considered that the <u>soil microbiological equilibrium</u>
either not at all or very slightly changes. In order to illustra-
te this fact, the counts (in log term) of different soil micro-
organism groups in soil receiving mineral fertilizer or a
heavy dose (SS_5) of sewage sludge are presented in figure 1.
On the basis of counts, bacteria dominate in the soil which are
followed by actinomycetes, algae, hyphal fungi and yeasts. This
is independent of the fertilizer treatments. The ratio between
different individual microbiological groups is quite constant
with an exception of algae. In no case, the effect of sewage
slugde application has a stimulating or lowering effect on the
microorganisms counts to an order of magnitude. The differences
in counts between the soil samples of different years are
greater than the differences in case of fertilizer treatments.
This is distinctly observable in case of actinomycetes counts.
This means that soil microbiological equilibrium is primarily
affected by the factors which lie out side the domain of direct
influences even true for extreme fertilizer doses. The dominant
factors are, moreover, the soil properties (i.e. pH, organic
matter, texture), cultivation practices and weather conditions.

3.2.2. <u>Soil biological activity and decomposition processes</u>

The sewage sludge application increased the soil biological activity as indicated by <u>catalase activity</u> (table 7), which corro-
borates the observations made by various scientists (9, 19, 21,
22). The increase in the activity depended upon the applied
amounts of sewage sludge.

The increase in the activity of soils is also shown by enhanced
<u>soil respiration</u> (11, 17, 21, 22). In compare to mineral fer-
tilizer, the normal and heavy doses of sewage sludge increased
up to 94 % and up to 238 % respectively, the mineralization of
organic carbon measured as amount of CO_2 evolved. The unfertilized soil (control) usually also showed about the same high
amount of CO_2-evolution intensity as the soil receiving mineral
fertilizer. This can be granted that the increase in respiration
in plots receiving sewage sludge was largely due to the decomposition of organic carbon compounds present in the sewage sludge.

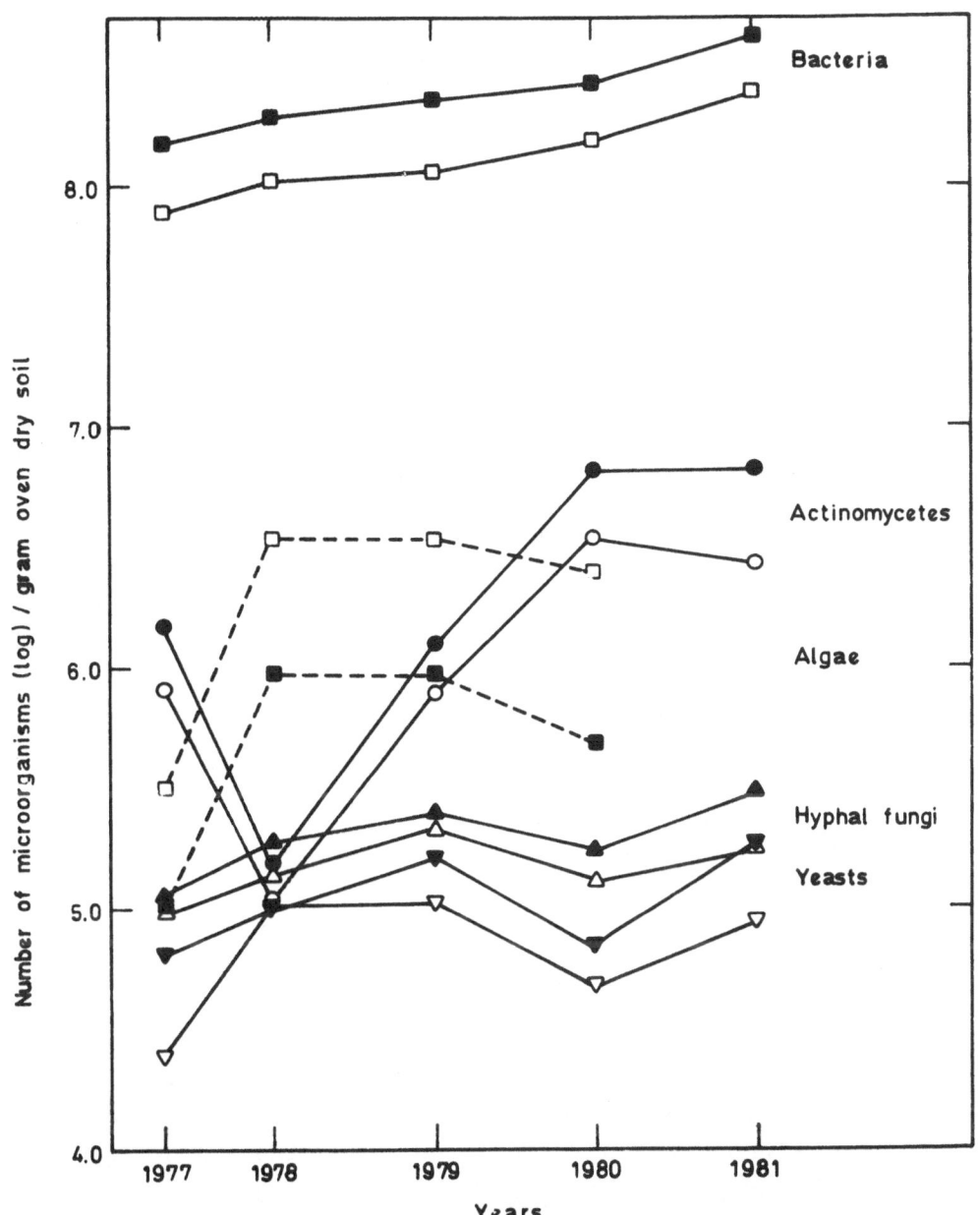

Figure 1 : The effect of heavy dose of sewage sludge on the different groups of microorganisms in a grassland soil compared to a mineral fertilizer (Average values of 4 replications).

Closed symboles = SS_5 = Sewage sludge heavy dose
Open symboles = Min = Mineral fertilizer.

TABLE 7 : INFLUENCE OF SEWAGE SLUDGE APPLICATION ON THE CATALASE
ACTIVITY, SOIL RESPIRATION AND AMMONIFICATION IN A
PERMANENT GRASSLAND SOIL. AVERAGE OF 4 REPLICATIONS
TREATMENT^{-1} YEAR^{-1}. (Min = Mineral fertilizer
treatment = 100).

Year	Treatment	Catalase	Respiration	Ammonification
1977	0	83	80	102
	SS_2	100	110	116
	SS_5	125*	172	137*
1978	0	100	85	126
	SS_2	126	134	106
	SS_5	137	187*	114*
1979	0	128	126*	132
	SS_2	129	150*	101
	SS_5	164*	193*	98
1980	0	105	119	97
	SS_2	117	194*	108
	SS_5	149*	338*	113
1981	0	108	102	104
	SS_2	119	143	105
	SS_5	168*	203 *	120*

Symbols used as shown in table 6.

Through up to 70 % of the total nitrogen associated with the sewage   sludge were present in organic form, the nitrogen supply
capacity of sludge treated soils, that means the ammonification,
in the spring samples of 1978-1981 was only slightly (up to 8 %
in case of SS_2 and up to 20 % in case of SS_5) higher as compared
to soil samples received mineral fertilizer. It seems, that a
large proportion of the organic nitrogen associated with sewage
sludge was already mineralized in the vegetation period, which
is substantiated by the enhanced ammonification in soil samples
of summer 1977 (table 7). After winter (exception 1979) even then,

extra nitrogen supply capacity in soil which received sewage
sludge was observable.

The recorded enhancement of C and N mineralization rates and
also catalase activity of sludge amended soils from soil
fertility and plant nutrition point of view can be positively
evaluated.

3.3. Comparison between the soil microbiology of arable and
     permanent grassland

Some of the investigated soil microbiological parameters from
the surface soil samples of 0-15 cm, which are collected in
late autumn of 1980 from the arable land plots (crop rotation)
and from the grassland plots receiving mineral fertilizer (Min)
or heavy dose of sewage sludge (SS_5) are presented in figure 2.
The following are the important observations :

- In case of mineral fertilizer treatment (Min), the biological
  activities i.e. alkaline phosphatase, catalase, respiration
  and the bacterial counts were found to be higher in grassland
  compared to arable soil. In permanent grassland, the activity
  of the alkaline phosphatase was 8 times greater than the activity in arable land. This observation could be explained from
  the facts that there were higher amount of root residues,
  larger amounts of root-excretions and the enzyme accumulation
  in the permanent grassland soil.

- The bacterial and actinomycetes counts and the biological
  activities (catalase, soil respiration, ammonification, alcaline phosphatase) were significantly affected by heavy sludge
  dose which is true for arable and grassland soils . In arable
  soil the sludge application brought about an increase of 84 %
  for aerobic bacteria, of 71 % for actinomycetes, of 16 % for
  catalase, of 50 % for soil respiration, of 6 % for ammonification and of 313 % for alkaline phosphatase as compared to
  mineral fertilizer.

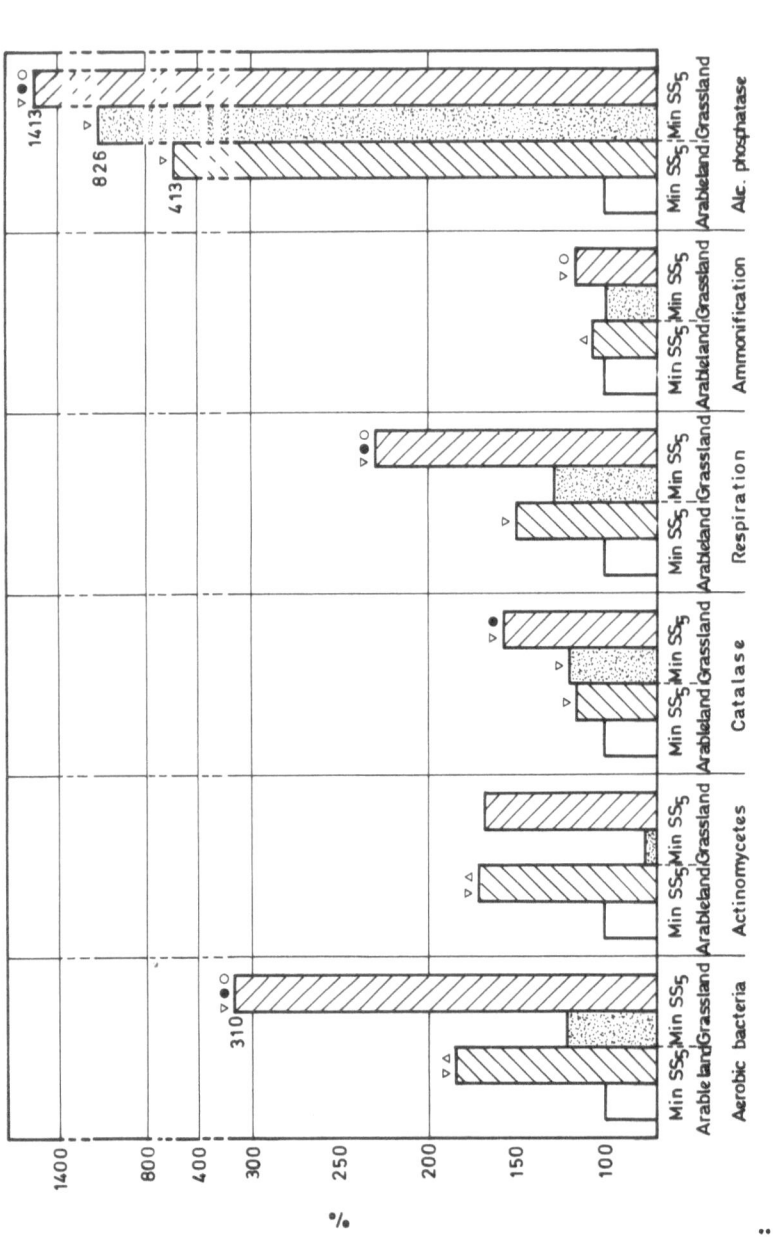

Figure 2 :

A comparison of arable and grassland soils, which received mineral fertilizer (Min) and a heavy dose of sewage sludge (SS_5), with respect to some of the microbiological parameters. (Arable soil with mineral fertilizer = 100%). Significant difference at 5% level (Wilcoxon-Range test) :

▼ other treatments   ＞ treatment Min arable land ● treatment SS_5 grassland ＞ treatment SS_5 arable land

△ treatment SS_5 arable land ＞ treatment Min grassland, ○ treatment SS_5 grassland ＞ treatment Min grassland

- With an exception of alcaline phosphatase, the sludge application had a greater effect on the soil microbial counts and
  their activities in the grassland soil than in the arable soil.
  In grassland soil the sludge application brought about an increase of 156 % for aerobic bacteria, of 199 % for actinomycetes, of 30 % for catalase, of 78 % for soil respiration, of
  18 % for ammonification and of 71 % for alcaline phosphatase
  as compared to mineral fertilizer. These results are supported
  from the observation that heavy doses of sewage sludge led to
  faster increase in humus content (table 4) in grassland soil
  than in arable soil (9).

- The lower activities which were regarded in arable soil compared to grassland soil receiving mineral fertilizer could
  partially be compensated by application of sewage sludge. This
  was true in case of bacterial counts and soil respiration but
  not in case of enzyme activities. Apparently the enzymes in
  arable land are less stable as compared to permanent grassland soil.

4. SUMMARY AND CONCLUSIONS

The influence of a normal dose and a heavy dose of sewage sludge
(2 t and 5 t organic matter ha^{-1} $year^{-1}$ respectively) on the
soil organic matter status and on few of the soil microbial parameters was studied in a field experiment started in 1976 on a
sandy loam soil. Both of these treatment were compared with
control plots (without fertilizer) and with a mineral fertilizer
treatment. Both a permanent artificial grassland and a three-
year-rotation (silage maize-winter wheat-clover grass) were
followed. Following points need consideration :

- In case of the heavy dose of sewage sludge, a slight increase
  in the organic substance ("humus" content) was recognizable
  in permanent grassland soil (0-20 cm) in contrast to arable
  soil after 4 1/2 years as compared to mineral fertilizer and
  the normal dose of sludge.

- The sludge application - in case of heavy doses more pronounced than in case of normal yearly doses - increased the
  most important heterotrophic soil microorganisms (aerobic
  bacteria, actinomycetes, yeasts, hyphal fungi), the mineralization processes (respiration, ammonification) and the biological activities (catalase, alcaline phosphate) but decreased the autotrophic algae.

- The sludge application had a greater effect on the soil microbial counts and their activities in the grassland soil than
  in the arable soil. The lower activities which are regarded
  in arable plots compared to grassland plots soils receiving
  mineral fertilizer could partially (e.g. in case of bacterial
  counts and respiration but not in case of enzyme activities)
  be compensated by the application of sewage sludge.

- The yearly overfertilization through sewage sludge for a
  period of four years has caused only slight changes (with an
  exception of soil algae) in the microbiological equilibrium of
  soil which are ecologically justifiable. The sewage sludge
  application left a positive balance on the soil microbiological activity, the biochemical processes in the soil environment, the soil fertility and the plant nutrition. Doubtful
  considerations from the soil microbiological point of view in
  connection with overfertilization of soil come from the facts
  that the soil microorganisms manage well with the over supply
  of sewage sludge, which could result for example in an increased NO_3 leaching, an increased N-loss through denitrification (5) and a disturbed biological nitrogen fixation.
  These phenomena must be clarified in future. In the long term
  point of view it must also be considered that heavy metals and
  other toxic substances of sewage sludge should pose a danger
  to the broad spectrum of soil microorganisms, to the biochemical processes of soils and to the microbial decomposing
  potential to eliminate foreign organic substances (24).

5. ACKNOWLEDGEMENTS

The authors express their thanks to Mr. V. Lehmann for looking after the field experiment, E. Fuhrimann, R. Schärer and R. Zimmermann for the soil microbiological determinations, F. Schär for humus determinations and S.K. Gupta for constructive suggestions.

6. LITERATURE

(1) AHRENS, E. : Beitrag zur Frage der Indikatorfunktion der Bodenmikroorganismen am Beispiel von drei verschiedenen Nutzungsstufen eines Sandbodens. Soil Biol. Biochem. 9, 185-191 (1977).

(2) BARTELS, R. : Der Einfluss von städtischen Abwasserschlämmen verschiedener Vorbehandlung auf die Bodeneigenschaften und die Ertragsleistungen von Sandböden. Diss. TU Berlin (1966), 109 pp.

(3) BECK, T. : Die Messung der Katalaseaktivität von Böden. Z. Pflanzenernähr. Bodenkd. 130, 68-81 (1971).

(4) BECK, T. Ueber die Eignung von Modellversuchen bei der Messung der bodenbiologischen Aktivität bon Böden. Bayer. Landw. Jb. 50 (3), 270-288 (1973).

(5) BECK, T. und SUESS, A. : Der Einfluss von Klärschlamm auf die mikrobielle Tätigkeit im Boden. Z. Pflanzenernähr. Bodenkd. 142 (2), 299-309 (1979).

(6) BOGUSLAWSKI, E. von : Die Verwertung von Klärschlamm verschiedener Aufbereitung im Ackerbau. Z. Acker- und Pflanzenbau 149, 406-423 (1980).

(7) BUCHNER, A. und STURM, H. : Gezielter düngen; intensivwirtschaftlich-umweltbezogen. Verlagsunion Agrar. DLG-Verlags-GmbH Frankfurt (1980), 319 pp.

(8) DIEZ, T. und WEIGELT, H. : Zur Düngewirkung von Müllkompost und Klärschlamm. Landwirtsch. Forschung $\underline{33}$ (1), 47-66 (1980).

(9) FURRER, O.J. : Einfluss hoher Gaben an Klärschlamm und Schweinegülle auf Pflanzenertrag und Bodeneigenschaften. Landwirtsch. Forschung Sonderheft $\underline{33}$/I, 249-256 (1977).

(10) FURRER, O.J. : Landwirtschaftlicher Wert des Klärschlamms. Landwirtschaftliche Verwertung von Abwasserschlämmen. EAS-Seminar in Basel vom 24.-26.9.1980.

(11) GLATHE, H. und MAKAWI, A.A.M.: Ueber die Wirkung von Klärschlamm auf Boden und Mikroorganismen. Z. Pflanzenernähr. Bodenkd. $\underline{101}$ (2), 109-121 (1963).

(12) GUPTA, S.K. : Ueber die Phosphat-Elimination in den Systemen H_3PO_4-γ-FeO(OH) und H_3PO_4 - $FeCl_3$ und die Eigenschaften von Klärschlamm-Phosphat. Diss. Bern (1976), 138 pp.

(13) HANSON, R.B. : Nitrogen fixation (acetylene reduction) in a salt marsh amended with sewage sludge and organic carbon and nitrogen compounds. Appl.Environ. Microbiol. $\underline{33}$ (4), 846-852 (1977).

(14) JAEGGI, W. : Die Bestimmung der CO_2-Bildung als Mass der bodenbiologischen Aktivität. Schweiz. landw. Forschung $\underline{15}$ (3/4), 371-380 (1976).

(15) MICHAEL, G., SCHMID, R. BARDTE, D., WILBERG, E. und MITSCHKE, C.: Auswirkung einer 10-jährigen Düngung mit verschiedenen Klärschlämmen auf Ertrag und Mineralstoffgehalt von Boden und Pflanzen. Landw. Forschung $\underline{33}$ (1), 38-46 (1980).

(16) MITCHELL , M.J. et al. : Effects of different sewage sludges on some chemical and biological characteristics of soil. J. Environ. Qual. $\underline{7}$ (4), 551-559 (1978).

(17) SCHAFFER, G. : Die Abwasserschlammverwertung auf landwirtschaftliche Nutzflächen. Z. Acker- und Pflanzenbau $\underline{126}$, 73-99 (1967).

(18) STADELMANN, F. : Der Einfluss von Klärschlamm auf die bio-
logische Aktivität und den Humushaushalt des Bodens. In-
formationstagung "Klärschlammverwertung in der Landwirt-
schaft". Schweiz. landw. Tech. Zollikofen, 24.3.1977,
pp. 27-43.

(19) STADELMANN, F. : Einfluss der Klärschlammdüngung auf die
Bodenmikroorganismen und deren Aktivität. COST Symposium
sur la caractérisation et l'utilisation des boues résiduaires Cadarache (France), 13.-15.2.1979, 9 pp.

(20) STADELMANN, F. und FURRER, O.J. : Die Wirkung steigender
Gaben von Klärschlamm und Schweinegülle in Feldversuchen.
II. Erste Resultate über die Auswirkungen auf Population
und Aktivität der Bodenmikroorganismen. Schweiz. landw.
Forschung (under publication).

(21) SUESS, A., ROSOPULO, A., BORCHERT, H., BECK, T.,
BAUCHHENSS, J. and SCHURMANN, J. : Experience with a pilot
plant for irradiation of sewage sludge. Results on the
effect of differently treated sewage sludge on plants and
soils. Int. Atom. Energy Vienna "Radiation for a clean
environment", 503-533 (1975).

(22) SUESS, A., T. BORCHERT, H. ROSOPULO, A., SCHURMANN, G. und
SOMMER, G. : Ergebnisse 3 jähriger Feldversuche mit unbe-
handeltem, pasteurisiertem und γ-bestrahltem Klärschlamm.
Bayer. Landw.Jb. 55 (4), 481-505 (1978).

(23) VOGEL, C. : Die landwirtschaftliche Verwertung von Abwasser-
klärschlamm auf einem mittelschweren Boden. Diss. Giessen
(1975) 172 pp.

(24) WALTER, C. et STADELMANN, F. : Influence du zinc et du cadmium sur les microorganismes ainsi que sur quelques pro-
cessus biochimiques du sol. Rech. agron. Sui 18 (4), 311-
324 (1979).

(25) WARING, S.A. and BREMMER, J.M. : Ammonium production in
soil under waterlogged conditions as an index of nitrogen
availability. Nature 201 (4922), 951-952 (1964).

(26) WERNER, D. : Stickstoff (N_2)-Fixierung und Produktions-
biologie. Angew. Botanik 54, 67-75 (1980).

DISCUSSION

Dr CATROUX :   Are the difference in bacterial counts statistically significant ?

Dr FURRER :   Yes, they are.

Dr CATROUX :   What's your conclusion in terms of soil fertility?

Dr FURRER :   It is difficult to translate the results of biological investigations in terms of soil fertility. Important is a certain equilibrium between different types of organisms. I mentioned
in my paper, the relationship bacteria/fungi
gives some indication on soil fertility. To much
fungi relative to the bacteria seems not so good
for soil fertility and indicates an acidification
of the soil.

Dr CATROUX :   Concerning the soil microflora, do you hope to
observe anything with non specific activities
like catalase, phosphatase and so ?

Dr FURRER :   An important result is that the biological
activity increases with the application of increasing doses of sludge and that also with high
doses a good humification takes place. In the
future we will do more specific investigations,
especially in connection with the nitrogen
metabolism.

Dr BERGLUND :   Are there any negative effects on plant production
from the decrease of algae ?

Dr FURRER :   Till now we did not observe negative effects
on plant production. But an effect of the di-
minished algal growth in sludge plots was found:
a Sharp decrease of epepedaphic species of
collembola like Isotomidae, Entomobyridae and

Sminthuridae. Algae are the most important part of the food of these species of Collembola (see the paper of Zettel).

Dr DANNEBERG : Is it possible to explain the decreasing number of algae as an effect of competition with the increasing number of heterothophic organism ?

Dr FURRER : We think the decreasing number of algae is mainly due to the dark colour of the soil surface of the sludge treated plots, introducing less light and higher temperature. We think also, that enough plant nutrients are present, and in this field there is no competition.

Dr BERGLUNG : Are any comparisons made regarding microbiological activities with high and low rates of heavy metals in the sludge ?

Dr FURRER : A such project is running in our Institute (Dr STADELMANN and MS. RUDAZ).

INFLUENCE OF SEWAGE SLUDGE APPLICATION ON MICROARTHROPODS
(COLLEMBOLA AND MITES) AND NEMATODES IN A SANDY LOAM SOIL

J. ZETTEL - J. KLINGLER

1. MICROARTHROPODS

From march 1979 to march 1980, soil samples were taken in monthly
intervals (depth 0-5 cm) and extracted with a modified TULLGREN-
apparatus. Collembola were determined to the family or genus,
mites to the suborders only. The interpretation of the collembola
data is impeded by a field effect the mite data are not yet tested.

2. COLLEMBOLA   (data collected by B. Bachmann)

Table 1 shows that euedaphic species like Onychiuridae (Onychiurus
armatus, Tullbergia krausbaueri) are not heavily influenced by
sewage sludge in comparison with the control, while all epedaphic
and hemiedaphic groups (most Entomobryidae, Isotomidae and
Sminthuridae, but some species being euedaphic) are distinctly
reduced. Only the Podurids (Hypogastrura assimilis, a
coprophilous species) are favoured, but much less than in the
plots with pig slurry application.

TABLE 1. RELATIVE ABUNDANCE (%) OF COLLEMBOLA IN CONTROL PLOTS
(C), PLOTS TREATED WITH MINERAL FERTILIZER (M), SEWAGE
SLUDGE (S, 5 t ORGANIC MATTER $ha^{-1}y^{-1}$) AND PIG SLURRY
(P, 5 t ORGANIC MATTER $ha^{-1}y^{-1}$)

	C	M	S	P
Poduridae	100	170	496	7 197
Onychiuridae	100	271	117	88
Isotomidae	100	81	27	45
Entomobryidae	100	92	67	88
Sminthuridae	100	99	27	105
TOTAL	100	116	55	134

TABLE 2. ABUNDANCE OF COLLEMBOLA AND POPULATION COMPOSITION

N= total number counted

	C		M		S		P	
	%	N	%	N	%	N	%	N
Poduridae	0.8	196	1.3	333	8.3	973	49.5	14106
Onychiuridae	15.8	3375	37.0	9157	33.8	3944	10.4	2957
Isotomidae	50.7	10824	35.3	8744	25.5	2978	17.3	4906
Entomobryidae	22.1	4719	17.5	4343	27.2	3187	14.6	4168
Sminthuridae	10.4	2220	8.9	2205	5.2	973	8.2	2342
TOTAL	100.0	21332	100.0	24782	100.0	11683	100.0	28479

The dominant families (see Table 2) were Isotomids in the C-plots,
Isotomids and Onychiurids in the M-plots and the Onychiurids in
S-plots. The trends for the two organic fertilizers (S, P) were
similar, but S showed lower densities (especially in Isotomids
and Sminthurids) except for Onychiurids.

3. MITES (data collected by G. Breuer)

Astigmata were represented only irregularly and in small numbers.
From Table 3 it is evident that sewage sludge application

extremely reduced Prostigmata, Astigmata and Cryptostigmata.

TABLE 3. RELATIVE ABUNDANCE (%) OF THE MITE SUBORDERS

	C	M	S	P
Mesostigmata	100	93.7	133.5	213.0
Prostigmata	100	90.5	42.0	79.5
Astigmata	100	56.0	25.7	206.3
Cryptostigmata	100	179.4	9.3	170.3
Total	100	129.2	35.2	132.9

In the plots C, M and P the rank order in dominance was Cryptostigmata - Prostigmata - Mesostigmata - Astigmata, in S-plots Prostigmata - Mesostigmata - Cryptostigmata - Astigmata. The increased number of Mesostigmata (Gamasids) is considered as an effect of the heavy increase of their prey, the podurid collembola (see Table 2).

As negative indicators for sewage sludge application, Cryptostigmata seem to be most suitable, as well as Sminthurids. Isotomids were very irregularly distributed, thus a negative effect of sewage sludge cannot be concluded yet and further investigations are needed. Yet it is not clear which component of sewage sludge is responsible for the fact that collembola populations are reduced by 45 % and mites by 65 %. But as euedaphic micro-arthropods are much less affected than epedaphic and hemiedaphic ones, it seems to be a component losing its toxicity when penetrating into the soil, probably some toxic anaerobic decomposition products.

TABLE 4. ABUNDANCE OF MITES AND POPULATION COMPOSITION

N = total number counted

	C		M		S		P	
	%	N	%	N	%	N	%	N
Mesostigmata	9.7	2238	7.1	2113	37.0	2993	15.7	4777
Prostigmata	44.5	10236	29.7	8811	50.6	4089	25.3	7731
Astigmata	0.8	175	0.4	98	0.6	45	1.2	361
Cryptostigmata	45.0	10376	62.8	18609	11.8	962	57.8	17653
Total	100.0	23026	100.0	29631	100.0	8089	100.0	30522

SOIL MICROBIAL ACTIVITIES AS AFFECTED BY APPLICATION OF COMPOSTED

SEWAGE SLUDGE

S. COPPOLA

1. INTRODUCTION

Within the agricultural utilization of sewage sludge, the pressure to reduce the costs of disposal by applying sludge at maximum rates can cause yield depression and high plant uptake of
toxic elements (Hucker, 1980).
These are of course the most important effects to avoid.
But heavy rates of application can also disturb other soil properties, among which biological ones are not to be regarded as
secondary in order to soil fertility. The influence of rates
and application frequency upon soil biological activities depends
on the type of sludge and soil, besides other factors, as climatic conditions, cultivation methods, and so on.
This report deals with the effect of different treatments with
composted sewage sludge on microbial activities of four types
of soil.

2. MATERIALS AND METHODS

Sludge compost was produced by the pilot plant of the Centre for
Industrial Microbiology of the University of Naples, by composting raw sewage sludge in mixture with wood-chips, through the
forced aeration system. Characteristics of the compost are
reported in the table 1.
The characteristics of soils assayed are listed in the table 2.
Soil A is a volcanic soil, developed on the yellow tuff of
Posillipo (Naples), fine-textured, with high percentage of
vitreous material and very low clay content. It is the typical

TABLE 1. CHARACTERISTICS OF COMPOSTED SEWAGE SLUDGE

Raw sewage sludge from mixed (residential + industrial catchments were partially dewatered by filter press (up to 10 per cent of dry matter), mixed with wood chips (sludge : wood chips ; 4:1 ; w.:w.) and composted by the forced aeration system (13 liters of air per cubic meter of material per min.) Yield amounted to 12.7 Kg of compost (45 % d.m.) from 100 kg of sewage sludge (5 % d.m.)

Dry matter ($g.Kg^{-1}$ wet material)		455
Organic matter ($g.Kg^{-1}$ dry matter)		500
Total nitrogen	(")	20
NO_3-Nitrogen	(")	0.255
NH_4^+-Nitrogen	(")	0.400
Phosphorus, P_2O_5	(")	17
Potassium, K_2O	(")	11
Calcium	(")	87
Magnesium	(")	7
Sodium	(")	7
Mercury ($mg.Kg^{-1}$ dry matter)		0.7
Copper	(")	195
Zinc	(")	2,000
Cadmium	(")	7
Chromium	(")	800
Lead	(")	185
Nickel	(")	38
Manganese	(")	300

TABLE 2 : STUDY OF THE INFLUENCE OF COMPOSTED SEWAGE SLUDGE UPON BIOLOGICAL PROPERTIES OF DIFFERENT SOILS CHARACTERISTICS OF THE SOILS

Soil sample	Skeleton	Coarse sand	Fine sand	Silt	Clay	Colour dry	Colour moist	Moisture pF 2.54	Moisture pF 4.2	Specific gravity g x cm³	Bulk density	Porosity	CaCO₃	Organic matter	Nitrogen	pH	C/N
A	7.0	30.6	45.5	13.2	10.7	2.5Y6/2	10YR3/2	20.8	9.0	2.34	0.95	59.4	--	1.67	0.11	7.4	8.8
B	3.0	20.0	33.9	27.4	18.7	10YR5/4	10YR2/2	32.9	18.2	2.22	0.90	59.4	--	2.96	0.17	7.2	10.1
C	1.0	2.7	26.6	24.8	45.9	2.5YR3/6	7.5YR3/2	27.8	21.1	2.38	1.09	54.2	--	3.35	0.18	7.4	10.8
D	-	11.1	27.4	28.3	33.2	10YR4/3	10YR3/4	26.9	17.4	2.34	1.09	53.4	--	2.06	0.14	7.3	8.6

soil of the N-NW area of Naples (Campi Flegrei), of the Islands
of Ischia and Procida. Similar soil occurs in some volcanic
zones of Lazio. Soil B is an alluvial sandy-loam soil, developed
on pyroclastic rocks. Rather poor in clay content, it is present in the Plan of Caserta. Soil C is a typical "terra rossa"
from Castellana (Province of Bari), iron-oxides rich, with high
percentage of not-expansible Kaolinite-type clay. "Terra rossa"
occurs in Italy in large areas of Puglie, Lazio, Venezia Giulia,
Toscana, Liguria, Sardegna. It also occurs in some areas of
Southern France, Spain, Greece, Yugoslavia, Algeria and Israel.
Soil D is a clay-loam soil from bauxite caves near Dragoni
(Province of Caserta).
The table reports data obtained by Buondonno and Buondonno
(1981).
The experimental method is summarized in the table 3. The moni-
toring of sludge mineralization has been carried out measuring
the CO_2 release by the barium peroxide method (Cornfield, 1961).
The same technique has been utilized to evaluate the ability of
control and sludged soil samples to mineralize organic carbon,
after addition of 5 per cent of wheat straw as carbon source,
5 mg per cent of ureic nitrogen and 5 mg of nitrate nitrogen.
The time-course of ammonification has been obtained by titrating
ammonia released by soil samples added of 5 g per cent of Bacto-
Casaminoacids, from Difco Lab., USA. Nitrification has been
simply studied after supply of 15 mg of NH_4^+-nitrogen per cent
grams of control and treated soil, then proceeding to regular
analyses of NH_4^+-nitrogen (Nessler reagent), NO_2^--nitrogen (Griess
reagent) and NO_3^--nitrogen (AgSO$_4$-phenoldisulphonic method).
Nitrogenase activity of soil samples has been assayed through the
technique of reduction of acetylene to ethylene (Stewart et
al., 1967). Measurements were carried out by gas-chromatographic
determinations (Fractovap Carlo Erba equiped with Porapak R
column), and elaborated according to Hardy and Holsten (1977).
All the above mentioned microbial activities have been studied
by incubation of samples at room temperature. Moreover, water
content of soils were carefully brought to 50 per cent of the
saturation value.
Viable counts of heterotrophic microorganisms were carried out
on soil-extract agar plates, inoculated with gradual suspensions

TABLE 3 : STUDY OF THE EFFECTS OF COMPOSTED SEWAGE SLUDGE ON
──────── SOME BIOLOGICAL SOIL PROPERTIES

EXPERIMENTAL STEPS

1. APPLICATION TO SOIL OF COMPOSTED SEWAGE SLUDGE
   Quantities (per cent, w.:w.):   0 - 1.25 - 2.50 - 5.00
2. MONITORING OF MINERALIZATION
3. (after 3 : 4 months)
   3.1. EVALUATION OF THE ABILITY OF CONTROL AND SLUDGED SOILS:
 	
      a. to mineralize organic carbon
      b. to ammonify organic nitrogen
      c. to oxidize NH_4^+-nitrogen
      d. to fix molecular nitrogen
 	
   3.2. COUNTS OF :
      a. total heterotrophic microorganisms
      b. ammonifiers
      c_1. NH_4^+-oxidizers
      c_2. NO_2^--oxidizers
      d_1. aerobic free-living nitrogen fixing microorganisms
      d_2. anaerobic and facultative free-living nitrogen
          fixing microorganisms.

of the samples. The most probable number of ammonifiers was determined according to Pochon and Tardieux (1962). Nitrifiers
were counted in liquid medium according to Coppier and De
Barjac (1952) as well as on double-layer-silica gel plates according to Soriano (1968). For nitrogen-fixing microorganisms
the medium of Line and Loutit (1971) was used for anaerobic and
facultative bacteria, the Hino and Wilson medium (1958) for
aerobic ones. This last medium was first used supplemented with
10 µg of nitrogen x ml^{-1} of casaminoacids. Then colonies were
transferred on the same medium, lacking the casaminoacids com-

ponent. The first results were refered to oligonitrophilic microorganisms. Only the strains able to grow on the second version of the nutrient medium were considered and assayed as authentic nitrogen-fixing organisms.

3. RESULTS AND DISCUSSION

The mineralization of composted sludge in the four soils is reported in the figures 1-4. The process appears regularly related to the C/N ratios of the soils, since biological activity results higher in the soil samples B and C than in A and D. Mineralization rates always depend on the amount of compost applicated and, as already noted in former observations from researches carried with soil A, only with the lowest treatments they follow a normal pattern, characterized by reduced increases after about one month of incubation (Coppola, 1980). The highest levels promote progressive increase, probably corresponding to overcoming adverse conditions able to lessen the rate of the transformation. At a similar extent, the completion of the process seems enough advanced after about three months, but only for the lowest applications.

The ability of control and treated soils to mineralize organic carbon is reported in the figures 5-8. The respiration of control soils after the amendment with wheat straw is comparable. The effect of sludgeing showes some benefits in the poorest soils (A and D). Anyway the heaviest application rates of composted sludge adversely affect this microbial activity. Microbial counts of heterotrophic microorganisms are reported in the table 4. Statistical analysis has shown not significant differences among the treatments; but the average comparison between control and treated soil samples have been generally resulted significant. Depressed activities recorded for some sludge treatments are therefore caused by chemical and/or physical conditions adversely affecting microbial metabolism.

Figure 1

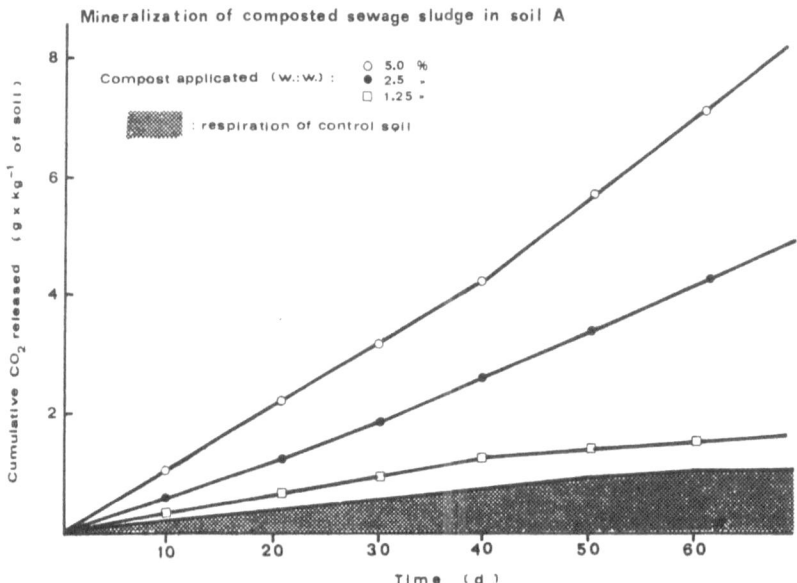

Figure 2

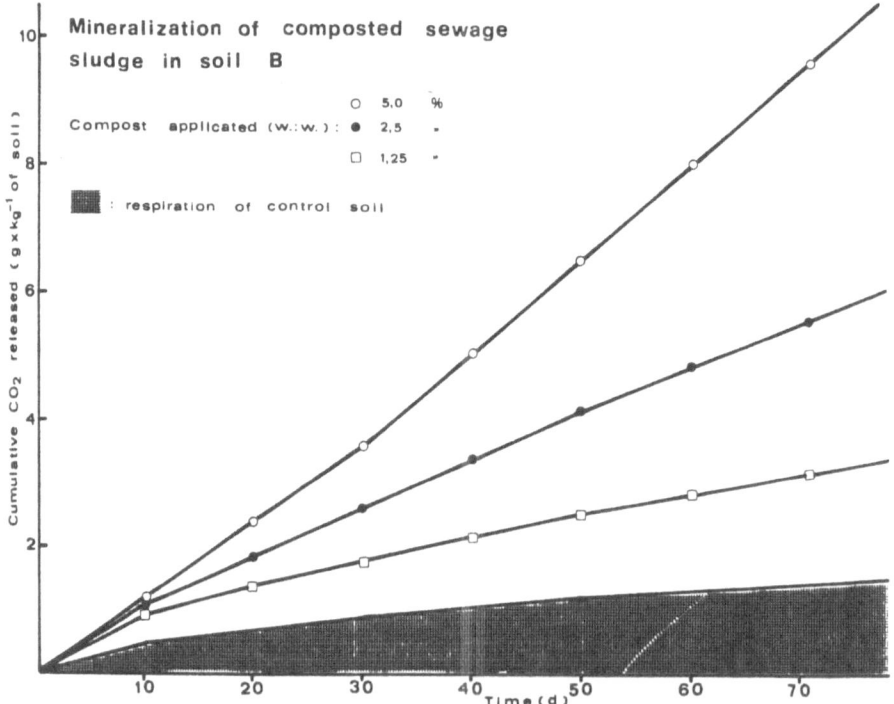

Figure 3

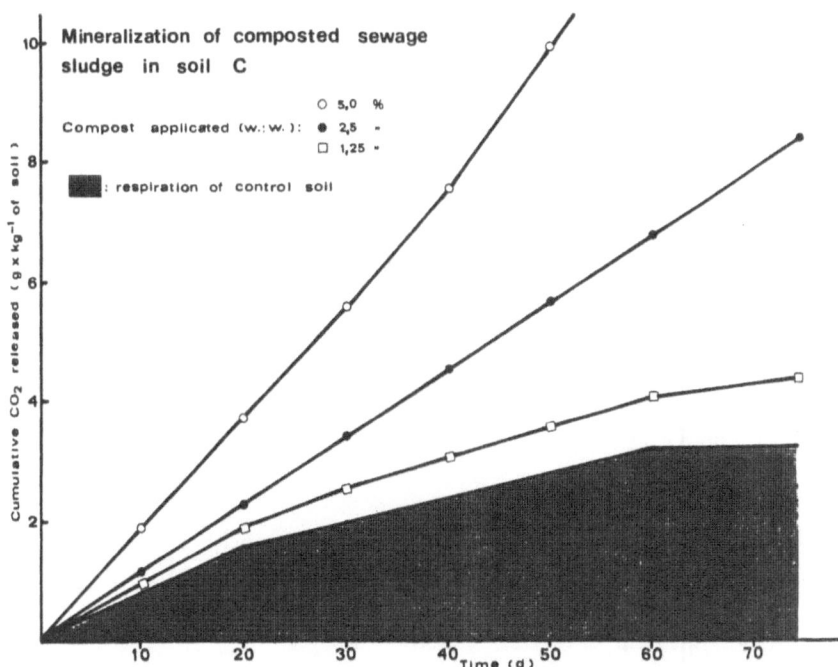

Figure 4

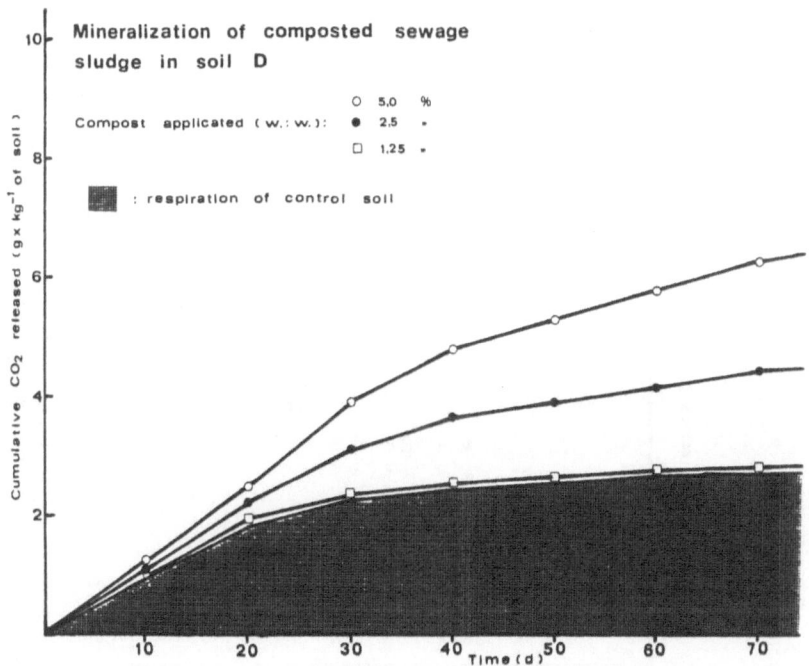

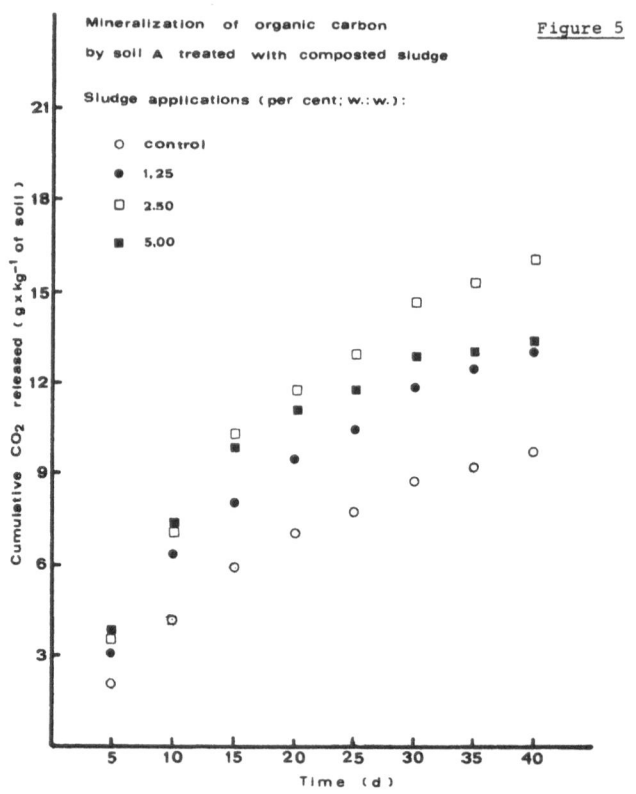

Mineralization of organic carbon by soil A treated with composted sludge

Sludge applications (per cent; w.:w.):

- ○ control
- ● 1,25
- □ 2,50
- ■ 5,00

Figure 5

Cumulative CO_2 released ($g \times kg^{-1}$ of soil)

Time (d)

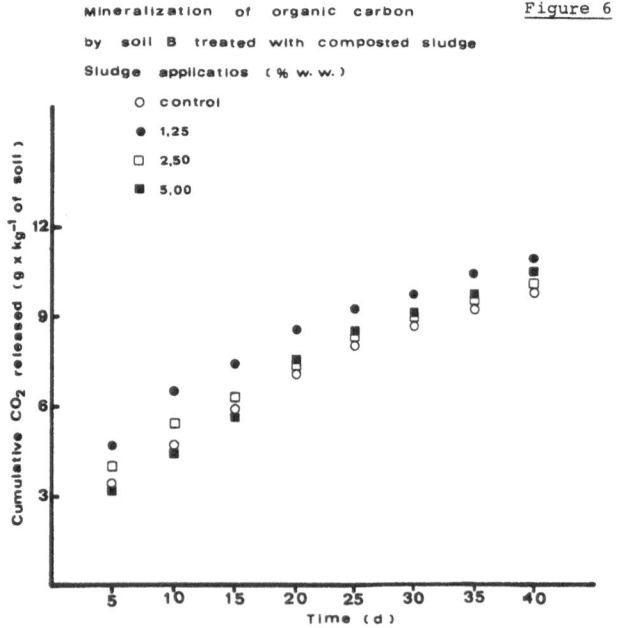

Mineralization of organic carbon by soil B treated with composted sludge

Sludge applicatios (% w.w.)

- ○ control
- ● 1,25
- □ 2,50
- ■ 5,00

Figure 6

Cumulative CO_2 released ($g \times kg^{-1}$ of soil)

Time (d)

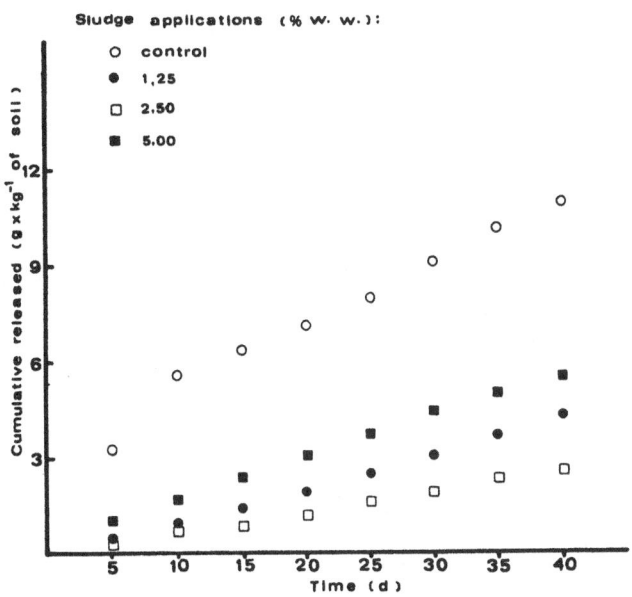

Mineralization of organic carbon by soil C treated with composted sludge — Figure 7

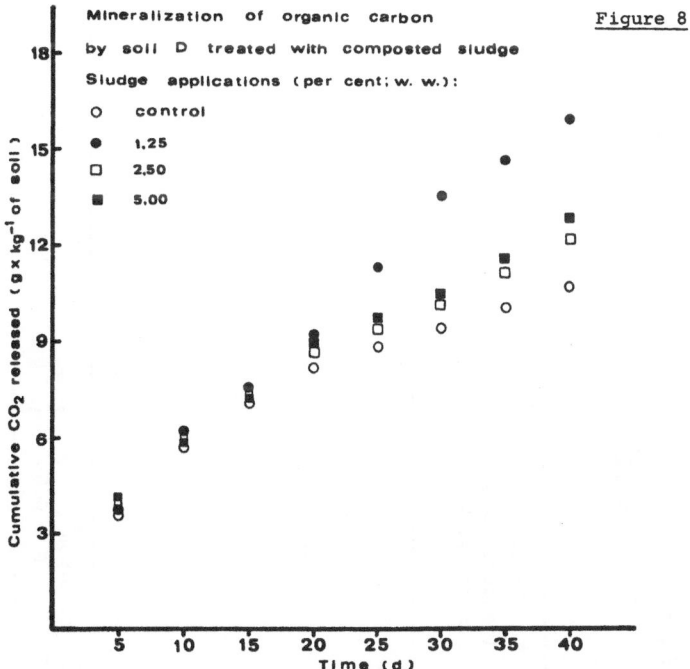

Mineralization of organic carbon by soil D treated with composted sludge — Figure 8

TABLE 4 : EFFECTS OF COMPOSTED SEWAGE SLUDGE ON HETEROTROPHIC MICROORGANISMS OF SOIL
(viable cells x g^{-1} d.w. of soil)

Treatments (percentage of compost added)	Soil samples			
	A	B	C	D
0	1.0×10^8	1.8×10^8	3.7×10^8	0.8×10^8
1.25	2.6×10^8	3.6×10^8	2.1×10^8	4.0×10^8
2.50	6.0×10^8	3.1×10^8	5.0×10^8	3.4×10^8
5.00	1.5×10^9	2.7×10^8	2.3×10^8	2.3×10^8
Significance F test; $P < 0.05$	$F = 2.94 < 3.10$	$F = 1.45 < 3.10$	$F = 0.25 < 8.53$	$F = 2.01 < 2.92$

Figures 9-12 report the effects of composted sludge on mineralization of organic nitrogen. The influence of sewage sludge
upon soil ammonification is quite negative at high application
rates, as resulted in all the four types of soil. However the
extent of such a influence is lower in the clay-soils. Moreover
ammonifier microorganisms are enough present in treated as well
as in control soil samples, as reported in the table 5. Therefore the decrease of microbial amino-hydrolase activity might
be attributed to organic and/or inorganic inhibitors that may be
partially neutralized by some soil components.

Nitrification rates vary from soil to soil : the sandy-loam soil
B showes the best activity, whereas the clay-loam soil D results
able to oxidize NH_4^+ at a lower extent. The figure 13 reports the
percentages of the different inorganic forms of nitrogen, after
two, four and six weeks of incubation of both control and treated
samples of soil. The figure points out the differences of nitrification among the four types of soil, but it also allows to remark that this microbial activity does not result compromised at
any extent by sludge treatments. Indeed composted sludge seems
generally to promote nitrification, probably by improving soil
structure. Counts of NH_4^+- and NO_2^--oxidizing microorganisms, reported in the tables 6 and 7, fundamentally reflect the activities
of the various soil samples and confirm the absence of any adverse influence.

The sensitivity of nitrifiers to heavy metals has been pointed out
out by many reports (Babich and Stotzky, 1978; Gupta and Haeni,
1980). Very probably, the amounts of toxic elements applicated to
soil by the heaviest treatments in these trials do not reach
effective levels. Alternately, the chemical forms of such metals
in the composted sludge could not be immediately active. The
influence of composted sludge upon dinitrogen fixation varies
with the soil sample. In the soil A the activity is enhanced at
a low extent, which appears directly proportional to the amount
of compost added. In the soil B the promotion of nitrogenase
activity by sludge amendments is higher. The behaviour of soil
C is not regular and only some application rates of compost show
to increase the process. In the soil D nitrogen fixation is

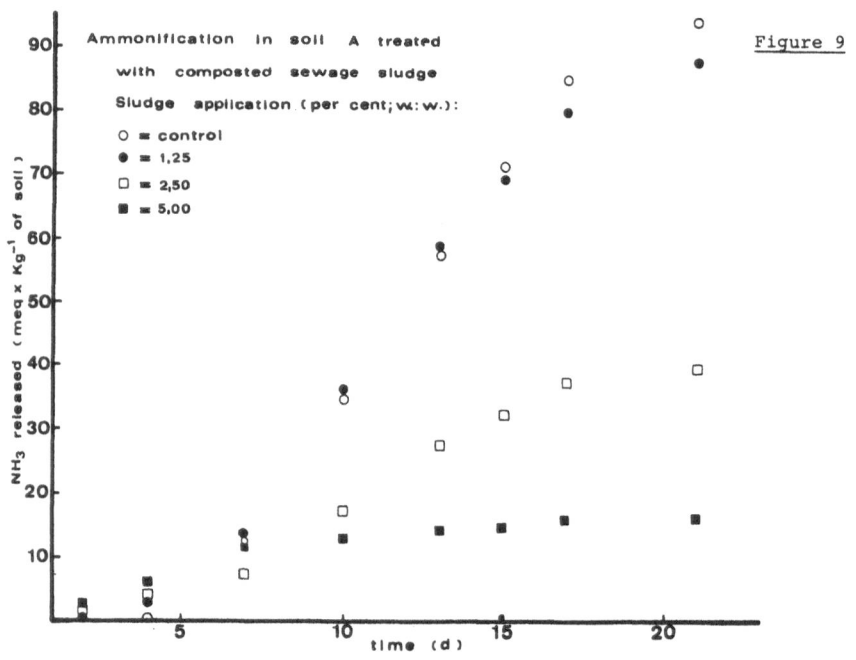

Figure 9

Ammonification in soil A treated with composted sewage sludge

Sludge application (per cent; w.:w.):

○ = control
● = 1,25
□ = 2,50
■ = 5,00

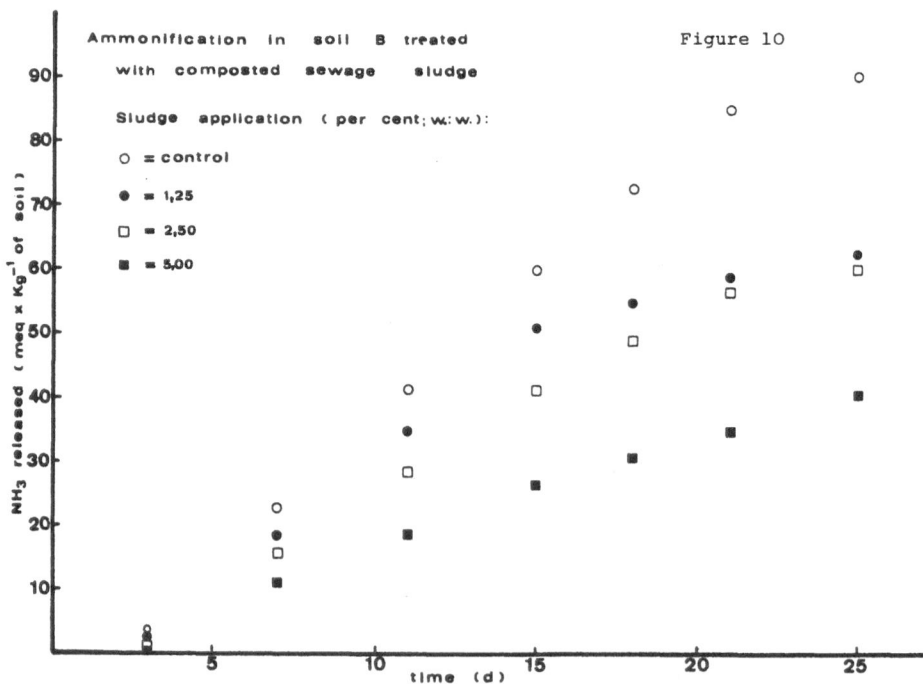

Figure 10

Ammonification in soil B treated with composted sewage sludge

Sludge application (per cent; w.:w.):

○ = control
● = 1,25
□ = 2,50
■ = 5,00

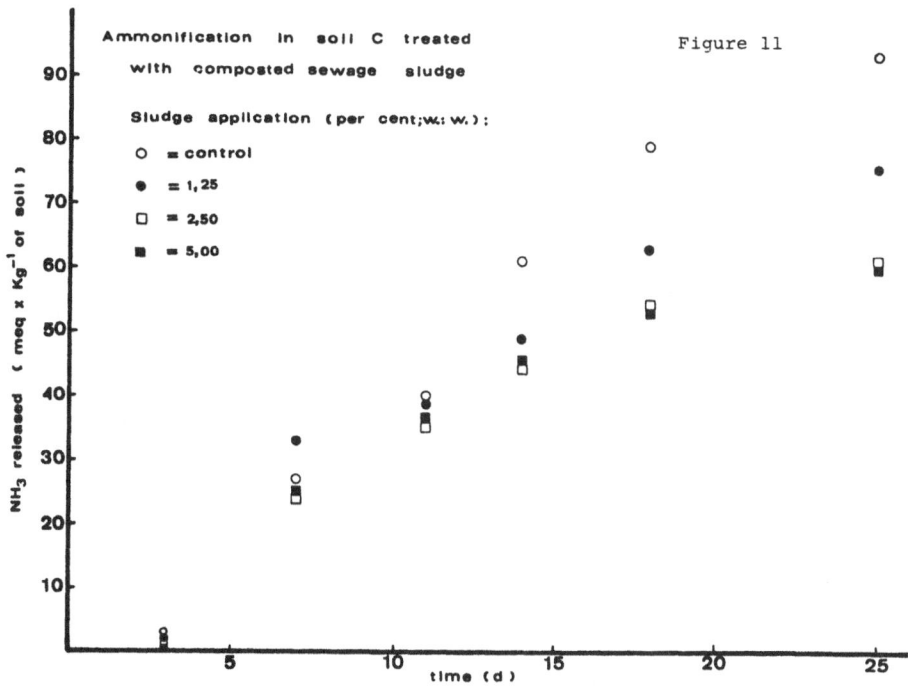

Figure 11

Ammonification in soil C treated with composted sewage sludge

Sludge application (per cent; w.: w.):

○ = control
● = 1,25
□ = 2,50
■ = 5,00

NH_3 released (meq x Kg^{-1} of soil)

time (d)

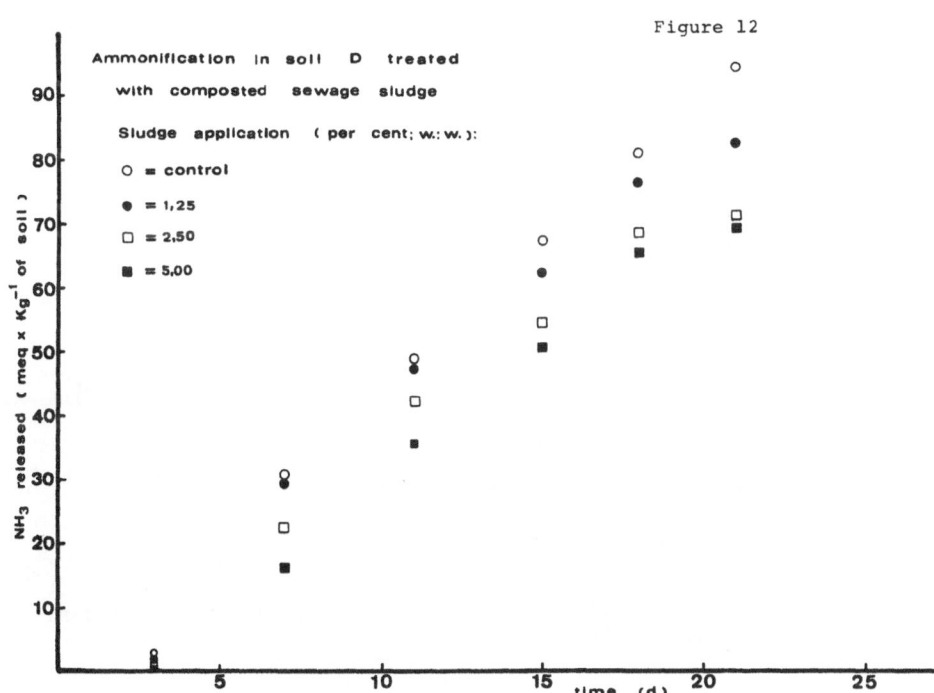

Figure 12

Ammonification in soil D treated with composted sewage sludge

Sludge application (per cent; w.: w.):

○ = control
● = 1,25
□ = 2,50
■ = 5,00

NH_3 released (meq x Kg^{-1} of soil)

time (d)

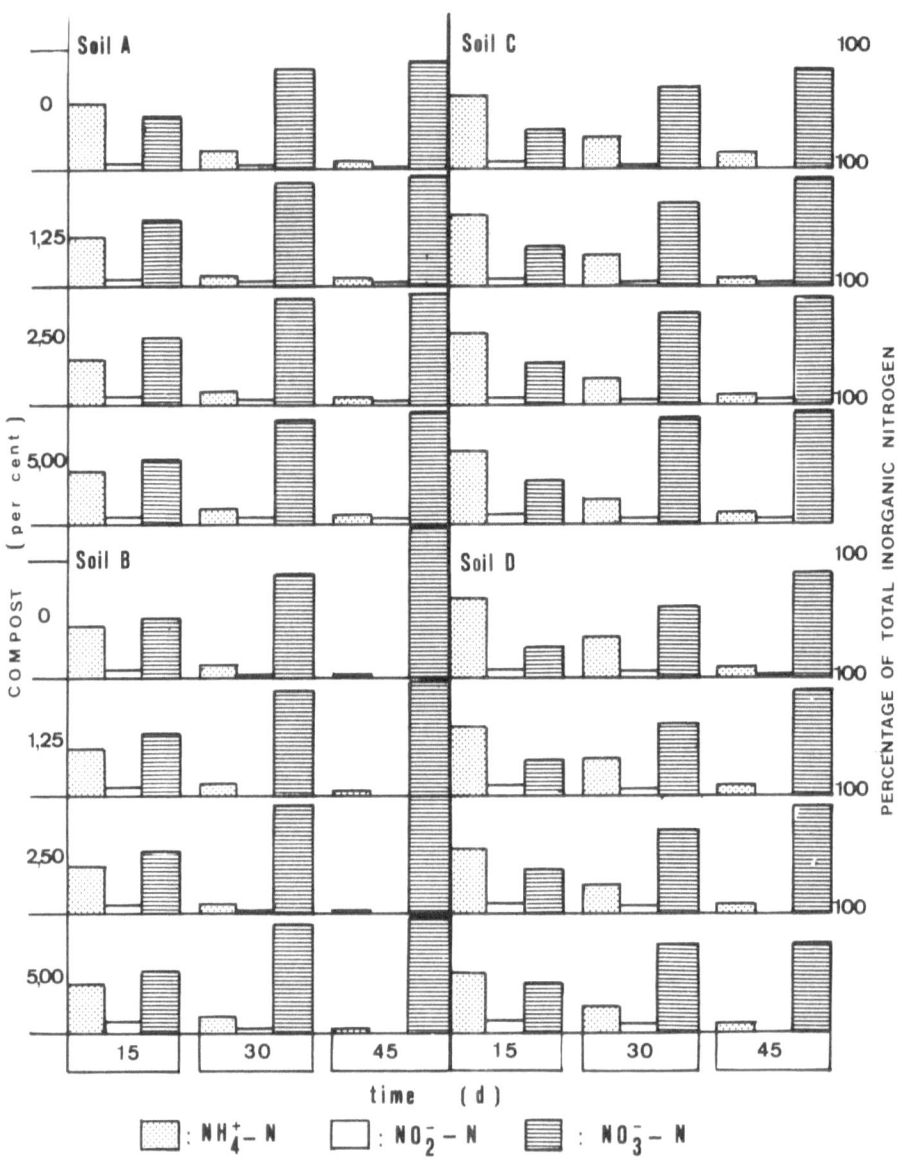

Figure 13 - Nitrification in soils treated with composted
sewage sludge

TABLE 5 : EFFECTS OF COMPOSTED SEWAGE SLUDGE ON AMMONIFIER MICROORGANISMS OF SOIL (MPN of viable cells x g^{-1} d.w.)

Treatments (percentage of compost added)	Soil samples			
	A	B	C	D
0	1.2×10^7	6.1×10^8	1.9×10^7	6.1×10^7
1.25	2.5×10^7	1.6×10^8	3.3×10^7	6.0×10^7
2.50	1.5×10^7	4.3×10^7	5.0×10^7	5.1×10^7
5.00	1.4×10^7	9.1×10^7	4.9×10^7	5.6×10^7

Table 6 - Effects of composed sewage sludge on NH_4^+- oxidizing microorganisms of soil (MPN of viable cells x g^{-1} d.w.)

Treatments (percentage of compost added)	Soil samples			
	A	B	C	D
0	1.1×10^4	3.4×10^4	6.1×10^2	2.0×10^3
1.25	1.2×10^4	1.2×10^5	1.2×10^3	1.2×10^3
2.50	1.0×10^4	3.5×10^4	1.3×10^3	3.4×10^3
5.00	9.6×10^3	2.1×10^4	1.3×10^3	3.0×10^3

Table 7 - Effects of composed sewage sludge on NO_2^-- oxidizing microorganisms of soil (MPN of viable cells x g^{-1} d.w.)

Treatments (percentage of compost added)	Soil samples			
	A	B	C	D
0	4.2×10^2	3.4×10^5	4.1×10^2	8.8×10^1
1.25	5.0×10^2	1.8×10^4	3.9×10^2	3.9×10^2
2.50	5.6×10^2	2.0×10^4	2.2×10^2	4.1×10^2
5.00	5.8×10^2	1.9×10^4	2.2×10^2	4.0×10^2

slightly depressed by compost treatment, but among the treated samples, the activity results proportional to the percentage of sludge applicated.

From the all four types of soil, count of oligonitrophilic microorgnaisms as high as 10^7 of viable cells per gram (3 x 10^6 in the soil A only) has been obtained . These are microbial strains able to grow on a low-nitrogen medium like as the Hino and Wilson agar supplemented with small quantities of casaminoacids and yeast extract. In effect the authentic nitrogen-fixing organisms, aerobic as well as anaerobic or facultative, are resulted present at a rather low extent (tables 8 and 9). However intense influences by sludge treatments upon the number of microorganisms belonging to these groups are not to enphasize. The sludge composted with wood-chips is a nitrogen-poor fertilizer. Moreover this compost is resulted to contain a very low amount of inorganic nitrogen. The application of such a material to soils with low organic content represents, almost exclusively, an energetic supply for nitrogen fixing organism. This can explain the beneficial effects pointed out at date. Therefore it seems interesting to continue the investigations in order to verify the influence exerted by further repeated applications.

In conclusion, apart some disturbances to mineralization processes, the application of composted sludge at high quantities too, does not appear to endanger at an important extent the microbiological activities assayed in four types of soil. The effects are more evident on microbial activities than on the number of the responsible microorganisms. Therefore only a chemical and/or physical influence upon microbial enzymes might explain the obtained results.

The heaviest compost application assayed within this experiment corresponds to about 2,000 quintals of composted sludge per hectar. Since higher single application does not seem suitable, we must conclude that only regular analyses of repeatedly sludged soils can assure the control of microbiological conditions.

TABLE 8 : EFFECTS OF COMPOSTED SEWAGE SLUDGE ON AEROBIC NITROGEN FIXING MICROORGANISMS OF SOIL
(MPN of viable cells x g^{-1} of soil d.w.)

Treatments (percentage of compost added)	Soil samples			
	A	B	C	D
0	4.16×10^2	1.98×10^2	1.96×10^1	9.92×10^1
1.25	4.86×10^1	0.86×10^1	1.64×10^1	2.10×10^2
2.50	1.55×10^2	1.56×10^1	0.86×10^1	5.32×10^1
5.00	2.10×10^2	0.90×10^1	0.88×10^1	2.02×10^1

TABLE 9 : EFFECTS OF COMPOSTED SEWAGE SLUDGE ON ANAEROBIC AND FACULTATIVE NITROGEN FIXING
MICROORGANISMS (MPN of viable cells x g^{-1} of soil d.w.)

Treatments (percentage of compost added)	Soil samples			
	A	B	C	D
0	8.80×10^4	2.05×10^6	8.56×10^5	4.67×10^5
1.25	9.92×10^3	9.86×10^5	4.96×10^5	1.63×10^6
2.50	4.57×10^4	9.88×10^5	6.25×10^5	1.38×10^6
5.00	8.05×10^4	1.49×10^6	1.06×10^5	1.35×10^6

4. ACKNOWLEDGEMENTS

These researches were supported by the Grant 80.01725.90 of Consiglio Nazionale delle Ricerche, Rome - Programme Finalizzato "Promozione della qualità dello ambiente". Author also thanks Mr. Alberto Cecere, Mr. Antonio Ausiello and Mr. Maurizio Pontonio, for their co-operation.

5. SUMMARY

Four different soils have been treated with increasing quantities of sewage sludge composted with wood chips by the forced aeration system.

The mineralization of sludge added to soil has been monitored during 100 days about. Then, some among the most important microbial activities have been evaluated in order to define the influence of frequency and level of compost applications.

Mineralization of organic carbon, ammonification, nitrification, nitrogen fixation as well as microorganisms responsible for these activities have been studied.

The importance of troubles caused by the highest application rates has been pointed out, and various responses have been regarded as depending on soil type.

6. LITERATURE

(1) BABICH, H. and G. STOTZKY (1978) Effects of Cadmium on the biota : Influence of environmental factors. Adv. in Appl. Microbiol., 23, 56-117.

(2) BUONDONNO, C. and A. BUONDONNO (1981). L'evoluzione degli sludges petroliferi nel suolo. Nota 1 : Prove preliminari in ambiente confinato. Liguori Editore, Napoli.

(3) COPPOLA S. (1980). Effect of composted sewage sludge on mineralization of organic carbon, ammonification and nitrification in soil. 2nd European Symposium on Characterization, Treatment and Use of Sewage Sludge, Vienna, 21st-23rd October.

(4) COPPIER, O. and H. DE BARJAC (1952). De la richesse d'un sol en microorganismes nitrificateurs. Ann. Inst. Pasteur, 99, 153-155.

(5) CORNFIELD, A.H. (1961). A simple technique for determining mineralization of carbon during incubation of soils treated with organic materials. Pl. Soil, 14, 90-93.

(6) GUPTA, S. and H. HAENI (1980). Easily extractable Cd-content of soil - Its extraction. Its relationships with the growth and root characteristics of test plants, and its effect on some of the soil microbiological parameters. 2nd European Symposium on Characterization, Treatment and Use of Sewage Sludge, 21st-23rd October.

(7) HARDY, W. and R.D. HOLSTEN (1977) in : Hardy, R.W.F. A treatise on dinitrogen fixation-Section IV : Agronomy and Ecology, by A.H. Gibson. John Wiley and Sons, New York.

(8) HINO, S. and P.W. WILSON (1958) Nitrogen fixation by a facultative bacillus. J. Bacteriol., 75, 403-408.

(9) HUCKER, T.W.G. (1980). Report of co-ordinator of Working Party Nr 5 on Second European Symposium on Characterization, Treatment and Use of Sewage Sludge, Vienna, 21st-23rd October.

(10) LINE, M.A. and M.W. LOUTIT (1971). Non-symbiotic nitrogen-fixing organisms from some New Zealand tussock grassland soils. J. Gen. Microbiol., 66, 309-318.

(11) POCHON, J. and P. TARDIEUx (1962). Techniques d'analyse en microbiologie du sol. Ed. de la Tourelle, St.-Mandé.

(12) SORIANO, S. (1968). Simplified methods for soil microbiological analyses. Rev. de Biol. du Sol., 9 , 7-8.

(13) STEWART, W.D.P., G.P. Fitzgerald and R.H. Burris (1967). In situ studies on N_2-fixation using the acetylene reduction technique. Proc. Natl. Acd. Sci. U.S.A. 58, 2071-2078.

DISCUSSION

Dr CATROUX : Have you made some extrapolations with your results on potential nitrogen fixation activities
to determine the amount of nitrogen possibly
fixed ? Is it sufficient to interest farmers ?

Dr COPPOLA : Conclusive results are not still available because statistical analysis of data are in progress.
My impression is that the improvement of nitrogen fixation due to composted sludge amendment
might be considered interesting for farmers,
mainly in poor soils.

Dr WILLIAMS : I am not clear on the conclusions which you have
drawn from your results ? Are you suggesting that
there is an immobilization of N immediately
following compost application for the first crop
grown and that a subsequent crop could benefit
as a result of N mobilization from the biomass ?

Dr. CATROUX : Concerning ammorifying activities, is it possible
to have mineral nitrogen immobilization due to
the composted sludge, during your incubation
test ? Can you really conclude to a detrimental
effect ?

Dr COPPOLA : Mineral nitrogen immobilization certainly occurs
during the incubation experiment for ammorifica-
tion. But I think that the extend at which such
a phenomenon might proceed in the experimental
conditions utilized, also considering the microbial counts found out in the samples and the
responses obtained by four different soils cannot
explain the intensity of the depression resulted
in soils treated with the highest dose of composted sludge. Any way, I feel that detrimental

action on mineralization processes are not to
be considered important with respect of long-
term period.

Dr CATROUX : Are you thinking it is really necessary to control the possible effects of sludges on microbial activities at farmers and field level ?

Dr COPPOLA : Heavy metals pollution of soils certainly must
be monitored within the practice of agricultural
utilization of sewage sludge. At a similar extent
metals contents of foodstuffs produced by sludged
soils are to be regularly examined . On the other
hand, scientific reports point out influences of
sewage sludge upon biological soil properties too.
Can we fully neglect such an effect on soil microorganisms and their activity ? In my opinion,
microbiological controls are necessary whenever
application rates exceed crop requirements. They
should at least cover the ability of treated soils
to mineralize organic matter, phosphatase activity, nitrification and nitrogen fixation. Every
sludge application should not find biological properties of soil at any extent compromised.

Dr. COPPOLA : Do we continue research on this topic ?

Dr DE HAAN : I am not a specialist in this field of research
and as far as I can see it there is not so much
need for research into the effect of sewage
sludge application on soil microbial activity
from a farmer's point of view, but the question
is put to us, not by farmers, but by environ-
mentalists, who are afraid that sewage sludge
application might disturb the activity of soil
organisms. We have to answer this question and
therefore research by specialists is necessary.

Dr CATROUX : Just a comment on the soil microbial counts.
M. COPPOLA products a table showing non statisti-

caly different heterotrophic bacterial counts.
However, results differ from near one Log. unit.
It's a general remark to inform the non-specia-
list on the weakness of soil microbiology method
and to the difficulty to observe differences
between treatments.

SIDE EFFECTS OF SEWAGE SLUDGES :

POSSIBLE ENHANCEMENT OF DENITRIFICATION

R. CHAUSSOD

1. INTRODUCTION

Field experiments, dealing with evaluation of nitrogen value of sewage sludges, take in account all the phenomena involved, including gazeous losses which are very difficult to estimate. It is therefore difficult to establish a nitrogen balance sufficiently accurate for a predictive use in other conditions (kind of sludge, soil type, climate...).

On the other hand, laboratory experiments are generally performed in standardized conditions where losses are impossible or -at least- minimized for example, aerobic conditions prevent losses through denitrification.

But it is quite evident that in wet soils denitrification may occur, and may be enhanced by landspreading of sewage sludge which are rather rich in organic carbon easily available for soil midroorganisms and used as electron acceptor in the denitrification process by denitrifying bacteria.

The laboratory experiments presented here were undertaken in order to point out the role of organic carbon brought by sewage sludges on denitrification.

2. MATERIAL AND METHODS

Measurement of denitrification

We used the acetylen method as described by GERMON (1980) from
the work of FEDOROVA (1973). FEDOROVA discovered that acetylen
stopped the reduction of nitrogen protoxyde N_2O to molecular
nitrogen N_2. The last step ($N_2O \rightarrow N_2$) of the denitrification
process (fig. 1) being blocked, nitrogen protoxyde is accumulated
in the flask during nitrate reduction. And N_2O is easily measured
by gas chromatography with a thermal conductivity detector, or
with electro capture for the low concentrations.

Soil, sludges and incubations

The loamy clay soil (org. matter 2 %) was air-dired and sieved
(2 mm) before utilization.

Two sludges were compared : a raw sludge (27.3 % C, 3.2 % N) and
an aerobic sludge (22.7 % C and 2.6 % N); the latter was obtained at the pilot scale from the former by COLIN (1978), so
they are strictly comparable and differ only from the stabilization treatment. The sludges were oven-dried (60°C) and ground
before utilization.

The incubations were conducted in 575 ml serum flaks containing
40 g (dry matter) of soil, either alone (control), or with
sludge added. In all cases, the sludges brought 8 mg total N per
flask, that is to say 200 mg N/kg soil.

The soil and the mixtures soil + sludge were moistened to reach
80 % of the field capacity and incubated at 20°C either 4h,
14 days, 28 days or 56 days according to the experiments. Carbon
dioxide evolved during this first (aerobic) incubation was trap-
ped in sodium hydroxide and recorded (fig. 2,a).

Figure 1 : DENITRIFICATION

$$NO_3 \longrightarrow NO_2 \longrightarrow NO \longrightarrow N_2O \xrightarrow{\quad\diagup\quad} N_2$$
$$C_2H_2$$

Figure 2 : INCUBATIONS

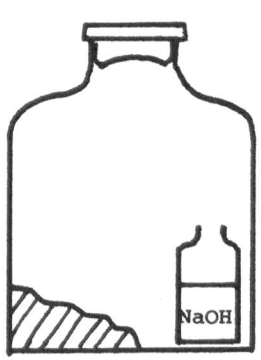

a : Aerobic incubation
   - 4 hours
   - 14, 28, 56 days

b : Flooding (t_0)
   - "anaerobic" incubation
     up to 356 h after
     flooding

Then, the soils were flooded with 50 ml water and acetylon
(2 %/vol) and krypton (as internal standard) were added (Fig.
2,b). This second incubation, mainly anaerobic, lasted up to
14 days, always at 20°C. Periodically, the gases in the atmosphere of the flaks were measured by gas chromatography : Kr,
CO_2, N_2O, C_2H_2. Only the values of N_2O are reported here; they
are corrected to take in account the variations of pressure in
the flaks and the solubility of this gas.

3. RESULTS

The following results are presented in two sets of experiments,
differing by the length of the first (aerobic) incubation :

- "short-term" experiments, after only 4 hours aerobic preincu-
  bation, with low and high nitrate levels.

- "long-term" experiments, with the nitrate content reached
  during a previous aerobic incubation.

A) "Short-term" experiments

In these experiments, the aerobic incubation lasted only 4 hours,
the minimum time to reach an equilibrium before flooding.

At the beginning of the experiment, the soil contained only
10 mg N/kg, and 24 to 36 hours after flooding all the nitrates
were reduced (lower part of the figure 3). If we add a nitrate
solution to reach 50 mg $N-NO_3$/kg soil, the complete reduction is
not achieved in the control after one week incubation. But with
sewage sludge, the rate of reduction is very important, pointing
out the enhancement of denitrification by sewage sludge (upper
part of the figure 3).

Of course, anaerobiosis is not reached immediatly after flooding
because the atmosphere and water still contain oxygen. But in
the soil, and especially with air-dried soil, enriched in avai-

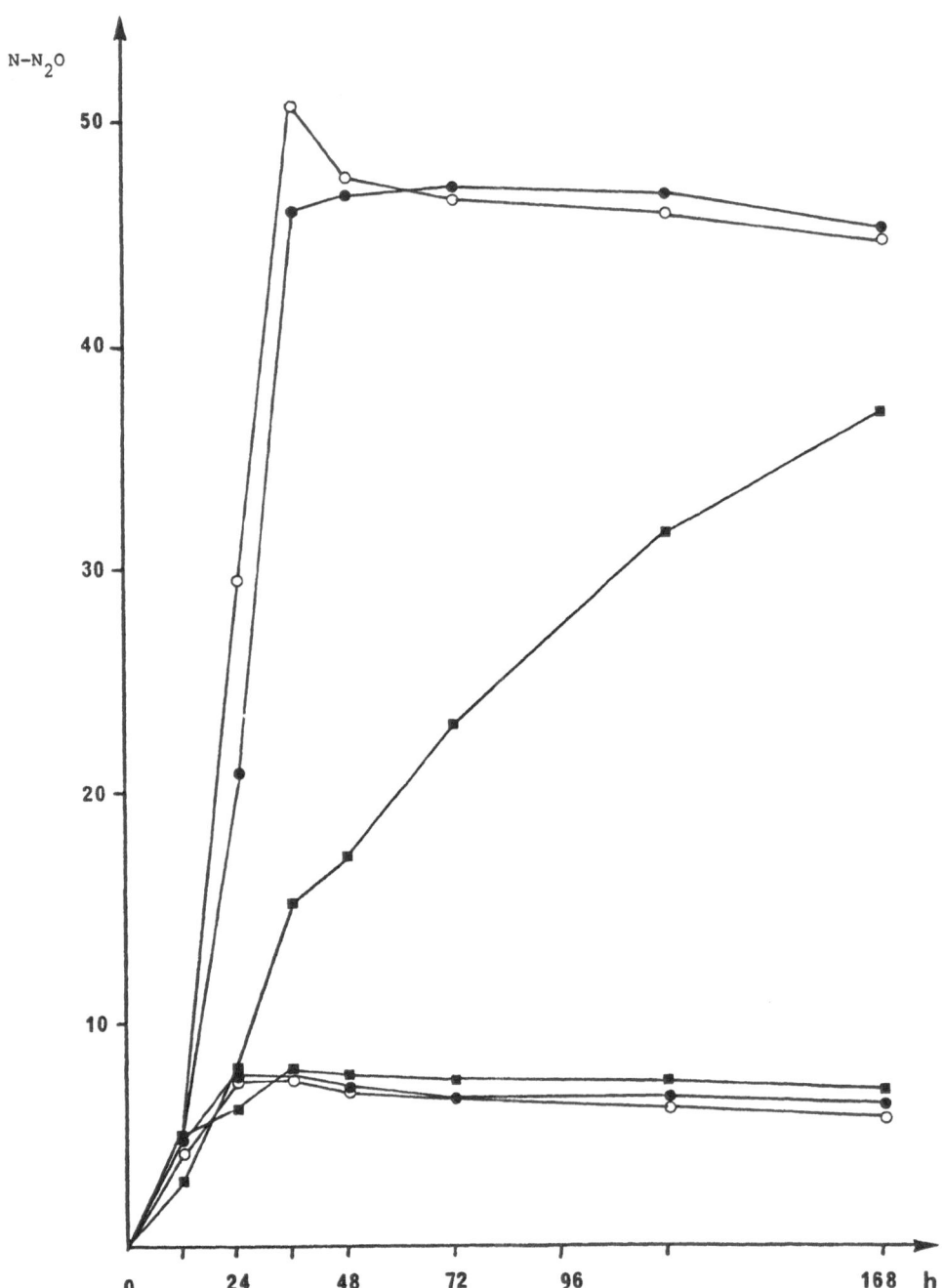

Figure 3 : Kinetics of N₂O production after flooding (mg N/kg soil)

■ Control (soil alone)
● Aerobic sludge
O Raw sludge

lable carbon, oxygen is consumed by microorganisms and the redox potential decreases sharply in the first hours. In the same way, the effect of sewage sludge is first to increase the oxygen consumption, leading quickly to anaerobiosis, then to stimulate denitrification through the organic carbon added.

B) "Long-term" experiments

In these experiments, the first aerobic incubation lasted 2, 4 and 8 weeks at 20°C. Meanwhile, a part of organic nitrogen from the sludges is mineralized and oxidized into nitrate; but a part of available carbon is simultaneously mineralized and therefore cannot participate in denitrification.

The kinetics of N_2O production are shown on figure 4. It is obvious that the losses of mineral nitrogen decrease with increasing the previous aerobic incubation. The nitrate losses are always much more important in the sewage sludges treatments than in the control (soil alone), and for each date, the denitrification is higher with the raw sludge than with the aerobic sludge. The nitrate content of the soil, before and after a 14 days anaerobic incubation, is given in table 1, where the losses are also expressed as a percentage of the nitrate content before flooding.

4. DISCUSSION AND CONCLUSION

In respect to the soil nitrogen cycle, sewage sludge are considered (CHAUSSOD, 1980) as an organic matter with carbon and nitrogen fractions of different availabilities for microorganisms. The fraction called "available carbon" is of great importance for the nitrogen value of sludges, either directly (nitrogen immobilization) or indirectly (denitrification).

In the experiments presented here, the amount of carbon evolved as CO_2 during the aerobic incubation is a part -and a good index- of "available carbon" ; it is about 20 to 30 % higher

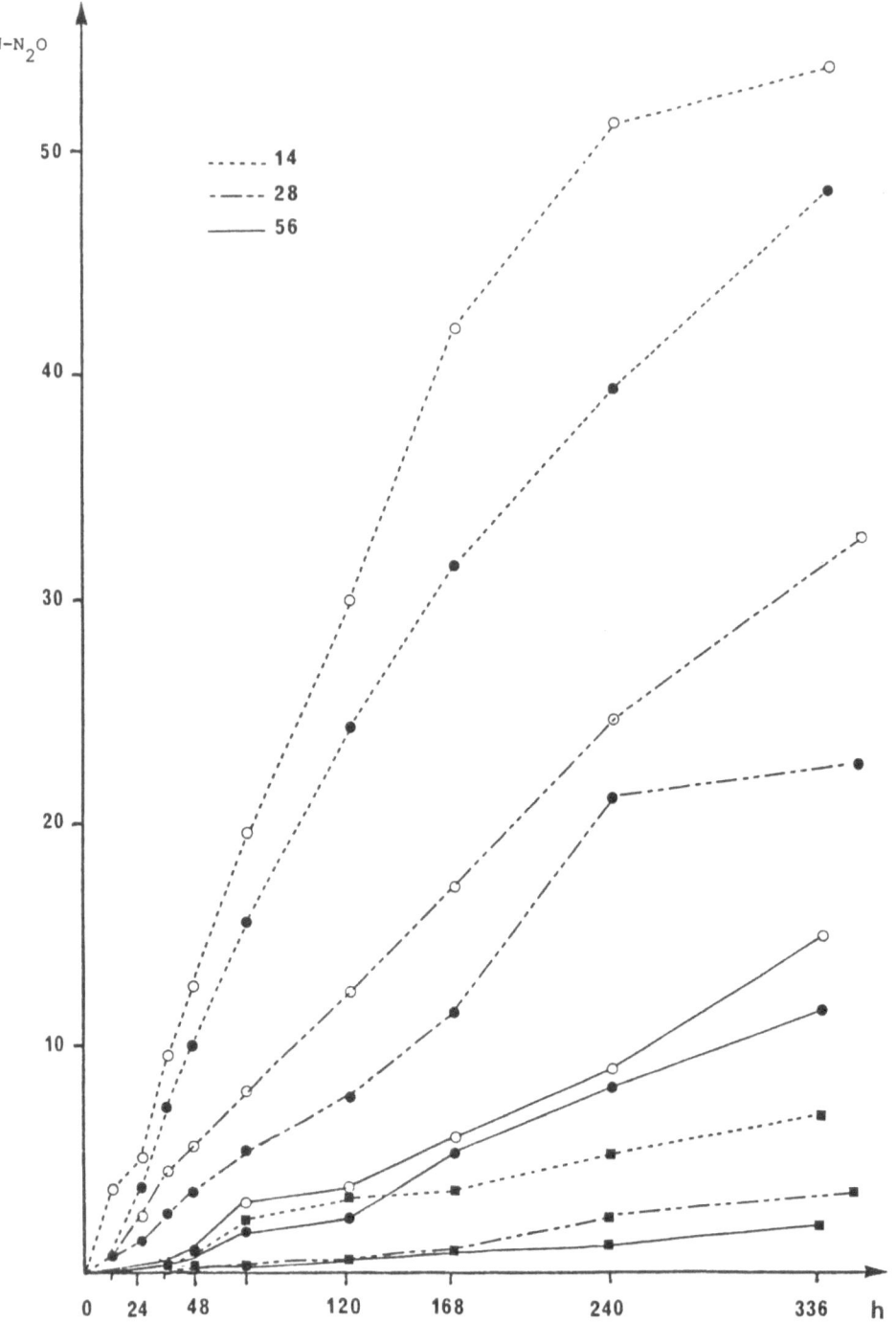

Figure 4 : Kinetics of N_2O production after three differnt previous aerobic incubations : 14, 28, 56 days.
(Symbols as in figure 3) Amounts in mg N/kg soil.

TABLE 1 : NITRATE BALANCE, VALUES ARE IN MG N/KG SOIL

	Aerobic incubation			Anaerobic incubation (14 days)		
Length (days)	Treatment	Evolved $C-CO_2$	$N-NO_3$ before flooding	$N-N_2O$ after 14 days	$N-NO_3$ remaining	reduced NO_3 (%)
14	control	165	26	6.8	22	26
	raw sludge	568	67	54	0	80
	aerobic sludge	435	64	48	10.4	75
28	control	250	34	3.6	31	10
	raw sludge	680	78	33	36	42
	aerobic sludge	539	79	25	38.5	28
56	control	313	41	2	38	5
	raw sludge	800	94	15	78	16
	aerobic sludge	645	89	12	74	13

for the raw sludge compared with the aerobic sludge (table 1),
and the losses through denitrification are always higher with
the unstabilized sludge. Moreover, the losses are not related to
the nitrate content of the soil, but to the available organic
carbon. So, in anaerobic conditions, sewage sludge greatly enhance denitrification, but the losses are somewhat lessened by
a good stabilization.

Even if the soil is in aerobic conditions, a heavy landspreading
of organic wastes such as more or less stabilized sludges may
lead to anaerobic conditions by increasing the oxygen consumption
in reduced gaseous exchanges with the atmosphere (wet soil for
example). A large part of the soil nitrate may be lost in the few
days or the few weeks following sludge application and especially
in the case of autumn application

5. SUMMARY

Laboratory experiments were performed to assess the possible
enhancement of biological denitrification in soil by an application of sewage sludge.

Under conditions of temperature and water content favourable to
denitrification, we recorded the nitrate reduction after application of two sludges differing mainly by their degree of stabili-
sation : a raw sludge and an aerobic sludge.

The nitrogen losses were compared to a control (soil without
sludge) by using the acetylen method : with 2 % C_2H_2 in the atmosphere, the reduced nitrate accumulate as N_2O, this gas being
measured by gas chromatography (thermal conductivity detector).

When an excess of water induced anaerobic conditions immediately
after the addition of sludge, the added nitrates disappeared
much more rapidly with the raw sludge than with the stabilized
sludge.

If an aerobic incubation took place before anaerobiosis, the quantity of volatilized nitrogen decreased for both sludges in relation to the length of previous (aerobic) incubation, i.e. in relation to the decrease of available organic carbon in spite of an increase of the nitrate concentration in the soil.

It can be concluded that the amount of readily available carbon (substrate for microbial growth and electron accepter) plays a major role in nitrogen losses by denitrification. Further more, anaerobiosis can be favourized by oxygen consumption during sludge decomposition in a wet soil.

6. LITERATURE CITED

(1) CHAUSSOD (R.), 1980 - Valeur fertilisante azotée des boues résiduaires. In : 2nd European Symposium on the Characterization, Treatment and Use of Sewage Sludge. Vienna 21-23 October 1980, p. 449-465.

(2) COLIN (F.), 1978 - Coordination des recherches sur les boues résiduaires. Juin 1978, document IRH (Nancy) 29 p.

(3) FEDOROVA (R.I.), MILEKHINA (E.I.) and ILYUKHINA (N.I.), 1973 - Evaluation of the method of "gas metabolism" for detecting extraterrestrial life. Identification of nitrogen - fixing microorganisms. Izv. Akad. Nauk. SSSR (Ser. Biol.), 6, p. 797-806.

(4) GERMON (J.C.), 1980 - Etude quantitative de la dénitrification biologique dans le sol à l'aide de l'acétylène. Ann. Microbiol. (Inst. Pasteur), 131 B, p. 69-90.

DISCUSSION

Dr DE HAAN :   In cultivated soils normally aerobic conditions prevail, but with heavy precipitation maybe for a period of 1-2 days anaerobic conditions might prevail. Do you think this period is long enough to bring about a measurable loss of nitrogen due to denitrification ?

Dr CHAUSSOD :   In these conditions, the losses of nitrogen are probably not very important. But, if the "available organic carbon" content is high (after landspreading of sludge, or manure, or another organic matter), the losses may be appreciable... according to the temperature because denitrification is highly sensitive to the temperature.

Dr DE HAAN :   Can there be a loss of nitrogen even under aerobic conditions in soils with much readily decomposable organic matter. In my experiments with addition of organic matter rich in nitrogen I have an indication that this is the case. The soils are kept under moisture conditions normal for aerobic circumstances.

Dr CHAUSSOD :   If the soil is really in aerobic conditions, denitrification cannot occur; in this case, nitrogen may be lost by ammonia volatilization. But, when highly decomposable organic matter is added to the soil, the oxygen consumption may decrease enough the redox potential to reach, in microsites, anaerobic conditions.

Dr DE HAAN :   You mentioned that there is not much cellulose and lignin-like organic matter in sewage sludges. But if there is no lignin-like OM in sewage sludges what is then the reason for its rather great stability when added to soils ?

Dr CHAUSSOD : From GALWARDI (University of Texas), the cellulose content of sewage sludge is about 3-7 % and the lignin content 6-9 %. There are also high quantities of grease and wax and other organic carbon (lignin - "like" O.M. ?) which biodegradation is very slow.

Dr DANNEBERG : A comment.
It seems to me not necessary to consider lignin as the only stable substance. Sludge has undergone an intensive microbial transformation with billions of organisms that live developed and died and decayed. We know from humic acid chemistry that high molecular and more stable substances are formed from decaying organisms.

LAND APPLICATION OF SLUDGE : EFFECTS ON SOIL MICROFLORA

A. PERA, M. GIOVANNETTI, G. VALLINI* AND M. DE BERTOLDI*

1. INTRODUCTION

Sewage sludge and solid urban waste compost application to crop land is becoming a common practice in agriculture. It has the double advantage of permitting the disposal of waste material and of utilizing cheap fertilizers (3,4,8). In spite of it, there are few reports on the effects of agricultural application of (solid and liquid) waste on soil microbial populations and particularly on rhizosphere microorganisms (9,15,16).

This research was carried out to verify possible changes of rhizosphere microbial population and particularly of the most important physiological groups of microorganisms contributing to biological fertility of soil. Besides, we studied the alterations of plant-microorganisms relationship, concerning mycorrhizal symbiosis. The experiments were carried out in open field with maize as test plants, in plots treated with different kinds of inorganic and organic fertilizers, including sewage sludge and compost deriving from uban waste.

2. MATERIALS AND METHODS

Soil and plant culture. Experiments were made utilizing maize plants, a grain hybrid class 400, in dry monoculture. Soil composition is reported in Table 1. Maize plants were seeded in

* Istituo di Microbiologia Agraria e Tecnica, Universita di Pisa, Pisa-Italy.

June 1979, straight after land applications of different fertilizers. The date of seeding was late because of continous rain during the month of May.

Treatments. 500 m^2 plots were employed for each treatment and treated with :

1. nothing (control)
2. mineral fertilizer
3. liquid sludge (aerobically stabilized)
4. liquid sludge (anaerobically digested)

TABLE 1 : SOIL COMPOSITION

ITEM	VALUE
pH	5.8
Organic matter	0.9 %
Total N	0.075 %
P_2O_5	trace
K_2O	1.31 %
Clay	10 %
Silt	14.1 %
Sand	75.9 %
Cationic exchange capacity	13.4 %

5. compost deriving from 60 % organic fraction of urban solid wastes mixed with 40 % liquid aerobic sludge.
6. compost deriving from 80 % organic fraction of urban wastes mixed with 20 % liquid anaerobic sludge.
7. manure.

Amounts of inorganic fertilizers and organic matter supplied in each treatment are reported in Table 2. Some chemical data of the organic fertilizers used are reported in Table 3. Root samples for microbial analysis (rhizosphere and mycorrhiza) were taken monthly starting after plant emergence.

Table 2 - Mineral fertilizers and organic matter (Kg/ha) supplied with the different treatments

Type of treatment	N (Kg/ha)	P_2O_5 (Kg/ha)	K_2O (Kg/ha)	Organic matter (Kg/ha)
Mineral fertilization	250	120	120	0
Aerobic sludge	1,563	1,500	303	20,600
Anaerobic slugde	1,312	1,462	292	24,573
Compost (aerobic sludge added)	1,312	847	356	23,220
Compost (anaerobic sludge added)	1,260	570	382	20,745
Manure	675	300	900	23,966

Table 3 - Some chemical data of the different fertilizers used

Fertilizer	Total Solids %	Organic Matter % d.w.	N % d.w.	P_2O_5 % d.w.	K_2O % d.w.
Aerobic sludge	1.6	55.04	4.17	4.17	0.81
Anaerobic sludge	3.5	65.53	3.50	3.90	0.78
Compost (aerobic s.)	41.44	61.92	3.50	2.26	0.95
Compost (anaerobic s.)	45.06	55.32	3.36	1.52	1.02
Manure	20.88	63.91	1.80	0.80	2.40

Rhizosphere analysis. In order to determine any difference in
the rhizosphere microoganisms among different treatments, 10
root apices, about 1 cm long, from each treatment were cut and
put in tubes with 10 ml distilled water. Tubes were shaken
for 2 min in a vortex. Aliquots of this suspension were taken
in order to count total numbers of aerobic bacteria, actinomycetes and fungi, following microbiological methods described
by Pochon and Tardieux (13). Besides, the main physiological
groups of microorganisms were analysed, concerned with carbon
and nitrogen cycles (14) again following the techniques described by Pochon and Tardieux (13).

The following microoogranisms were determined and counted :

Bacteria : proteolytic, ammonia producing, ammonia oxidizers,
nitrite oxidizers, nitrogen fixing.

Fungi : pectinolytic, amilolytic, cellulolytic.

After transferring aliquots of the suspension to tubes and
plates, 8 ml of the remaining suspension was dried at 90°C,
in order to relate the numbers of microbes to g dry soil.

Mycorhiza analysis 10 g roots were taken from each plant
sample, cleared in KOH at 90°C for 1 h and stained in trypan
blue in lactophenol (12).

Percentage of vesicular-arbuscular mycorhizal infections in the
roots were measured under a dissecting microscope, following
the grid-line intersect method (7). Endomycorhizal fungal spores
in the rhizosphere soil were extracted by the wet-sieving and
decanting method (6), and then counted under a dissecting
microscope.

3. RESULTS AND DISCUSSION

Rhizosphere microorganisms. Total numbers of bacteria, actinomycetes and fungi isolated from rhizosphere is reported in Fig.
1, 2 and 3. The different fertilizers used showed no significant
influence on the growth of rhizosphere mircroorganisms. Variations in number of these microorganisms seem to be more related
both to plant growth phases and to plant nutrient availability
than to the different organic and inorganic additions. Generally,
the lowest microbial content was observed in controls which
did not receive any fertilization; no actinomycetes were detected in this thesis.

As far as physiological groups are concerned proteolytic and
ammonia producing bacteria did not show any significant variation among the treatments, even if controls always presented the
lowest values (Fig. 4, 5).

The ammonia oxidizers were generally inhibited by the presence
of high levels of organic matter and they totally disappeared
four months after the treatments (Fig. 6). The analysis methods
used here were able to detect autotrophic ammonia oxidizer microorganisms only, and their disappearance probably coincided
with a contemporary growth of heterotrophic ammonia oxidizer
microorganisms (2,10). In fact wa can see in Fig. 7 that, though
the total disappearance of autotrophic ammonia oxidizers, nitrite oxidizer bacteria are massively detected up to the end of
the experiment.

Nitrogen fixing bacteria (aerobic and free living) were isolated
from 10^1 to 10^4 cells/g dry weight soil. The number of these
bacteria increased in the later stages of plant growth, and in
the experiments where organic fertilizers were used (Fig. 8).

The physiological activities of fungi tested (pectinolysis,
amilolysis and cellulolysis) showed variations which again seem
to be more related to the plant growth phases than to the dif-

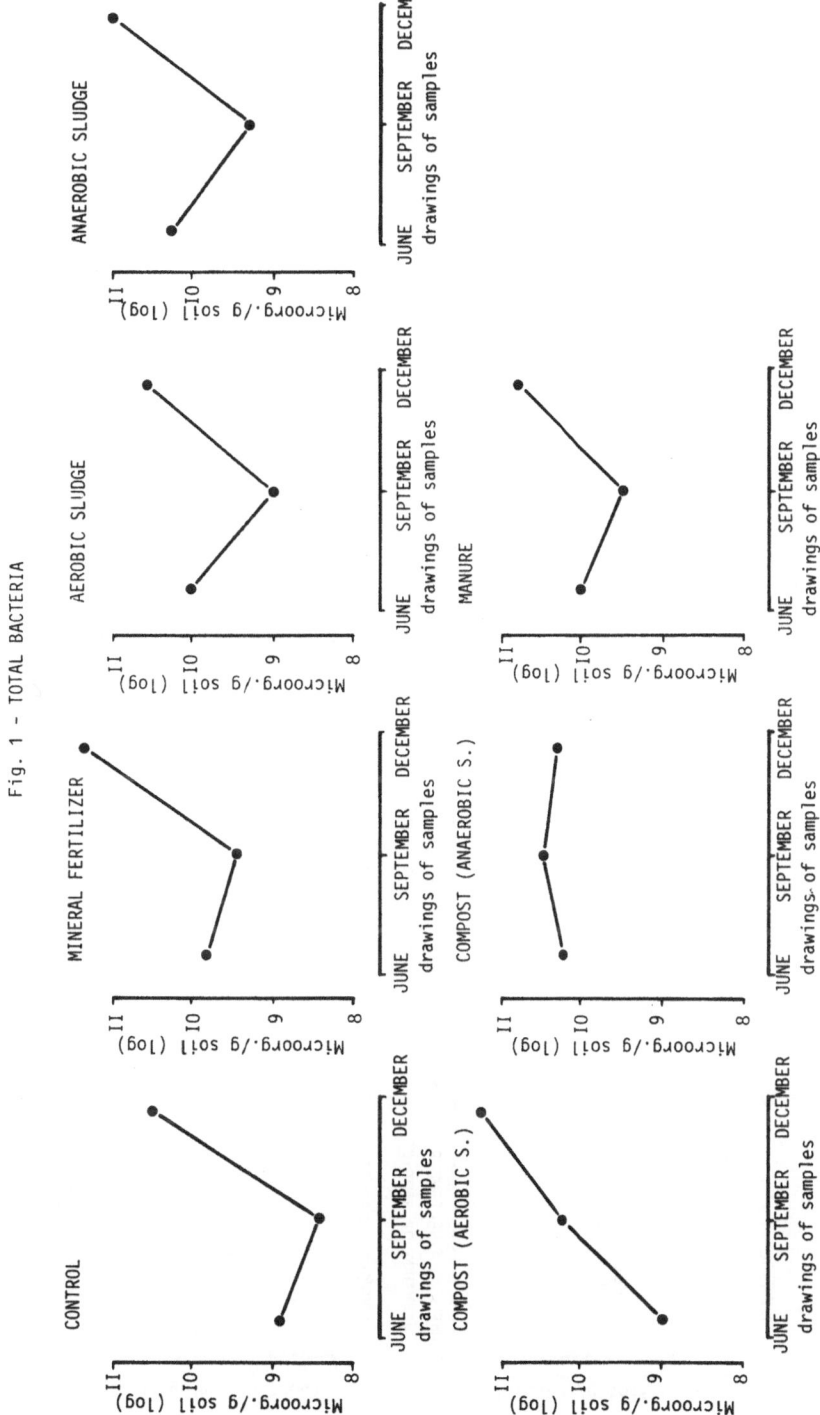

Fig. 1 - TOTAL BACTERIA

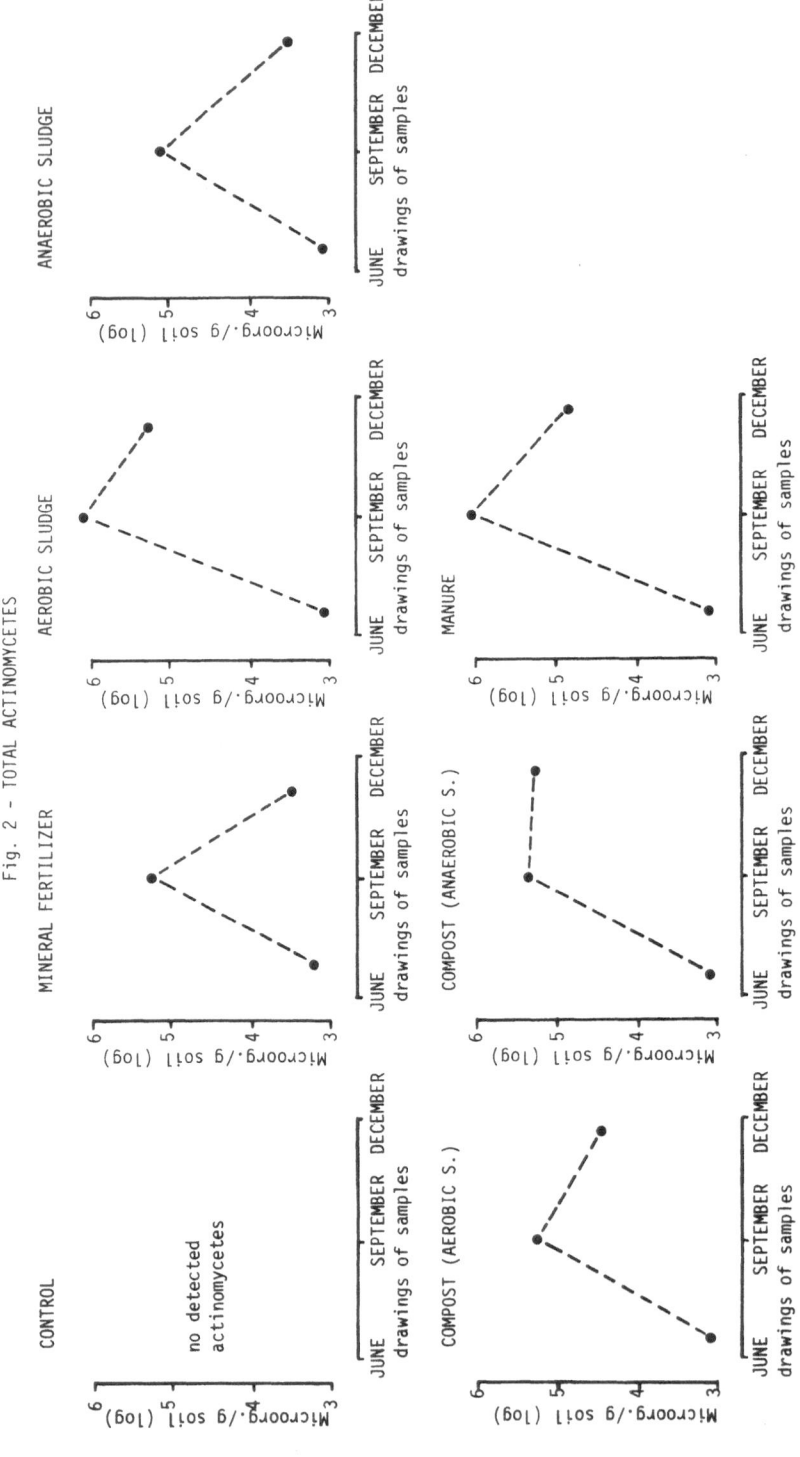

Fig. 2 - TOTAL ACTINOMYCETES

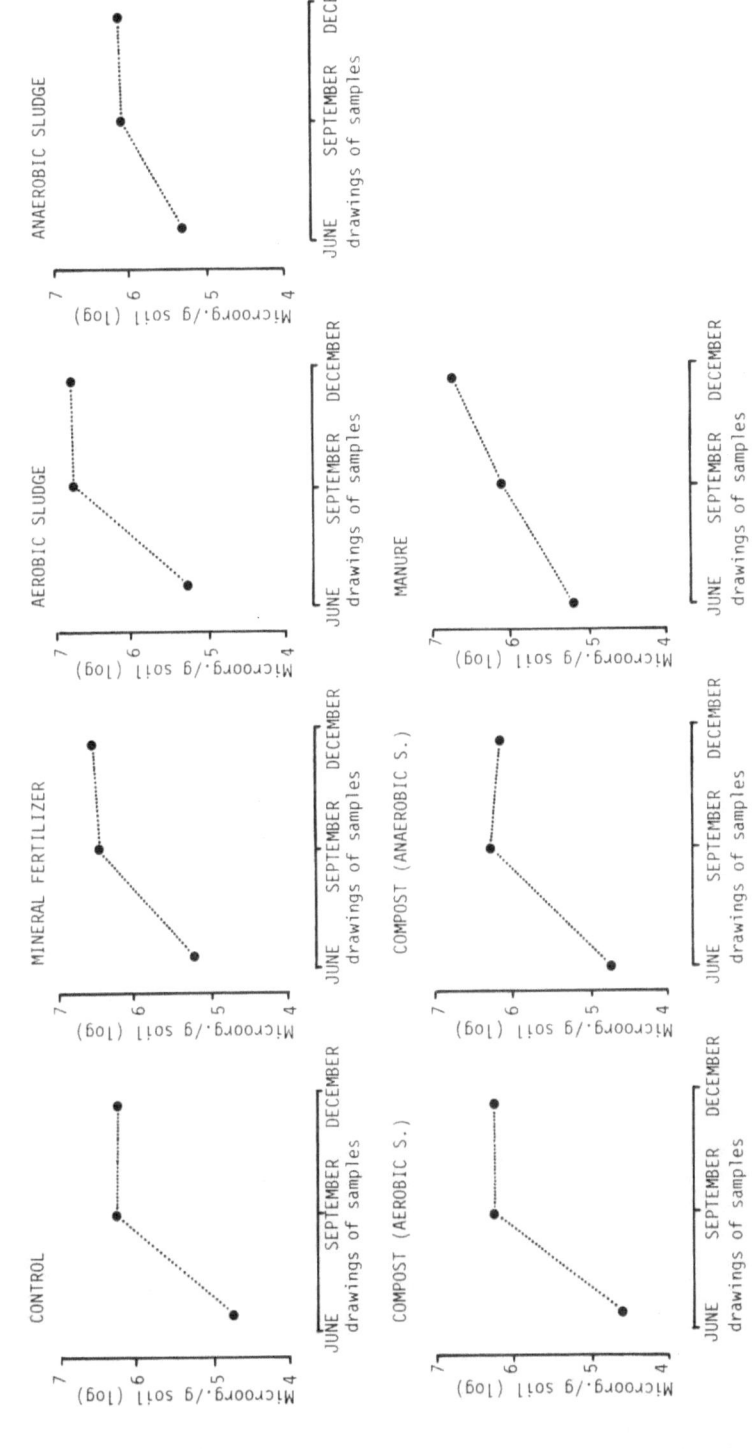

Fig. 3 - TOTAL FUNGI

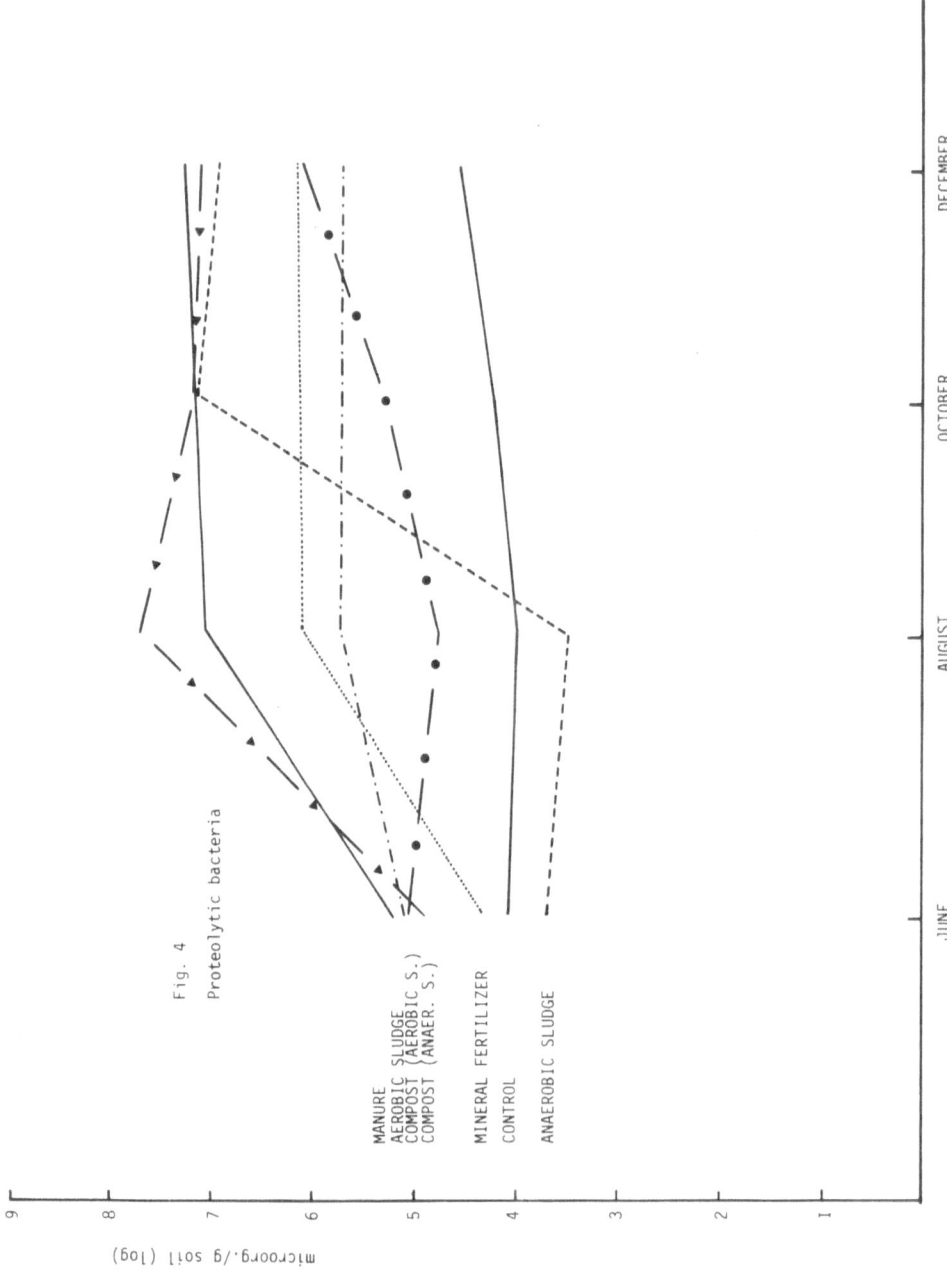

Fig. 4
Proteolytic bacteria

MANURE
AEROBIC SLUDGE
COMPOST (AEROBIC S.)
COMPOST (ANAER. S.)

MINERAL FERTILIZER

CONTROL

ANAEROBIC SLUDGE

microorg./g soil (log)

9
8
7
6
5
4
3
2
1

JUNE   AUGUST   OCTOBER   DECEMBER

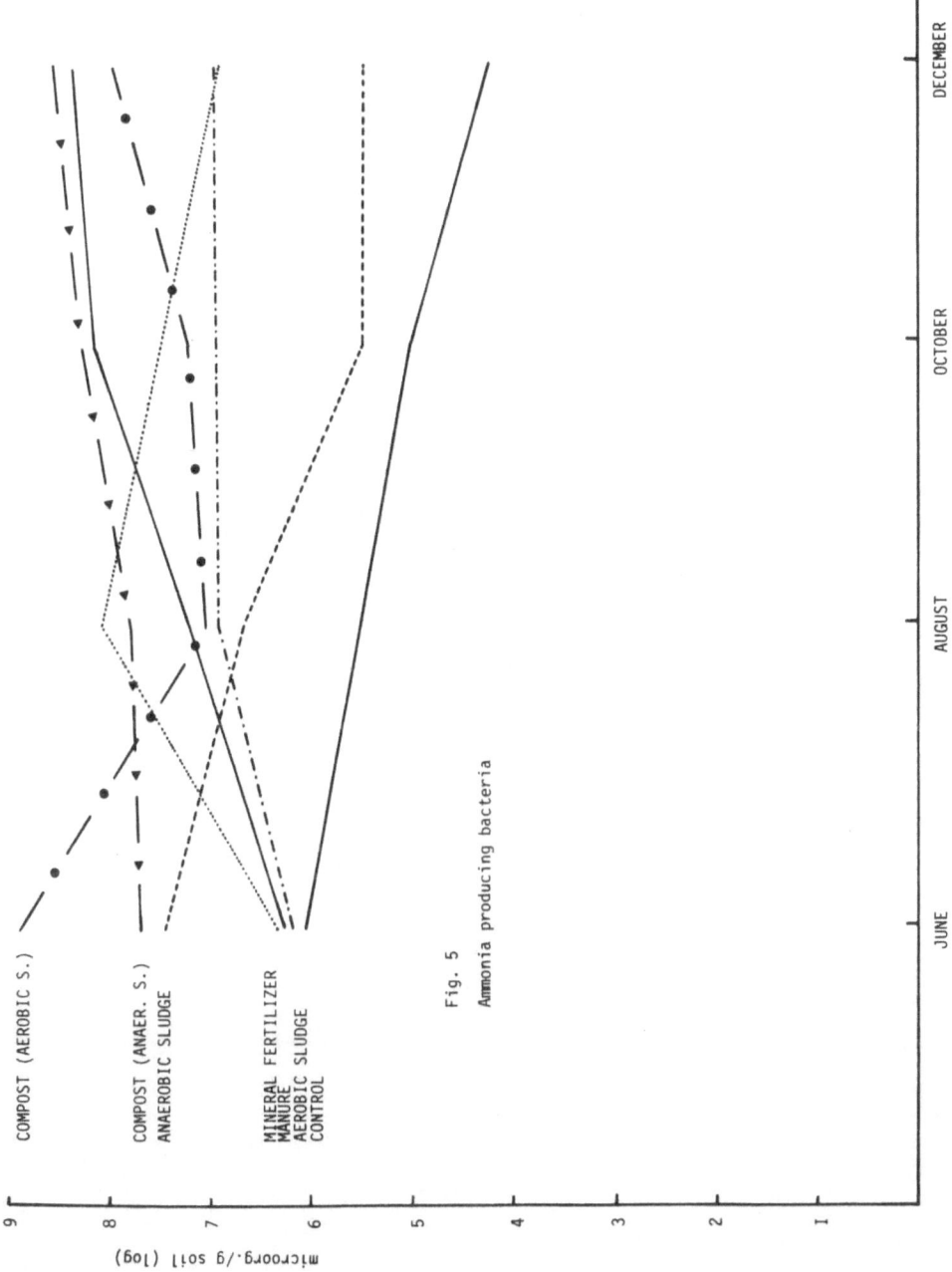

COMPOST (AEROBIC S.)

COMPOST (ANAER. S.)
ANAEROBIC SLUDGE

MINERAL FERTILIZER
MANURE
AEROBIC SLUDGE
CONTROL

microorg./g soil (log)

JUNE AUGUST OCTOBER DECEMBER

Fig. 5

Ammonia producing bacteria

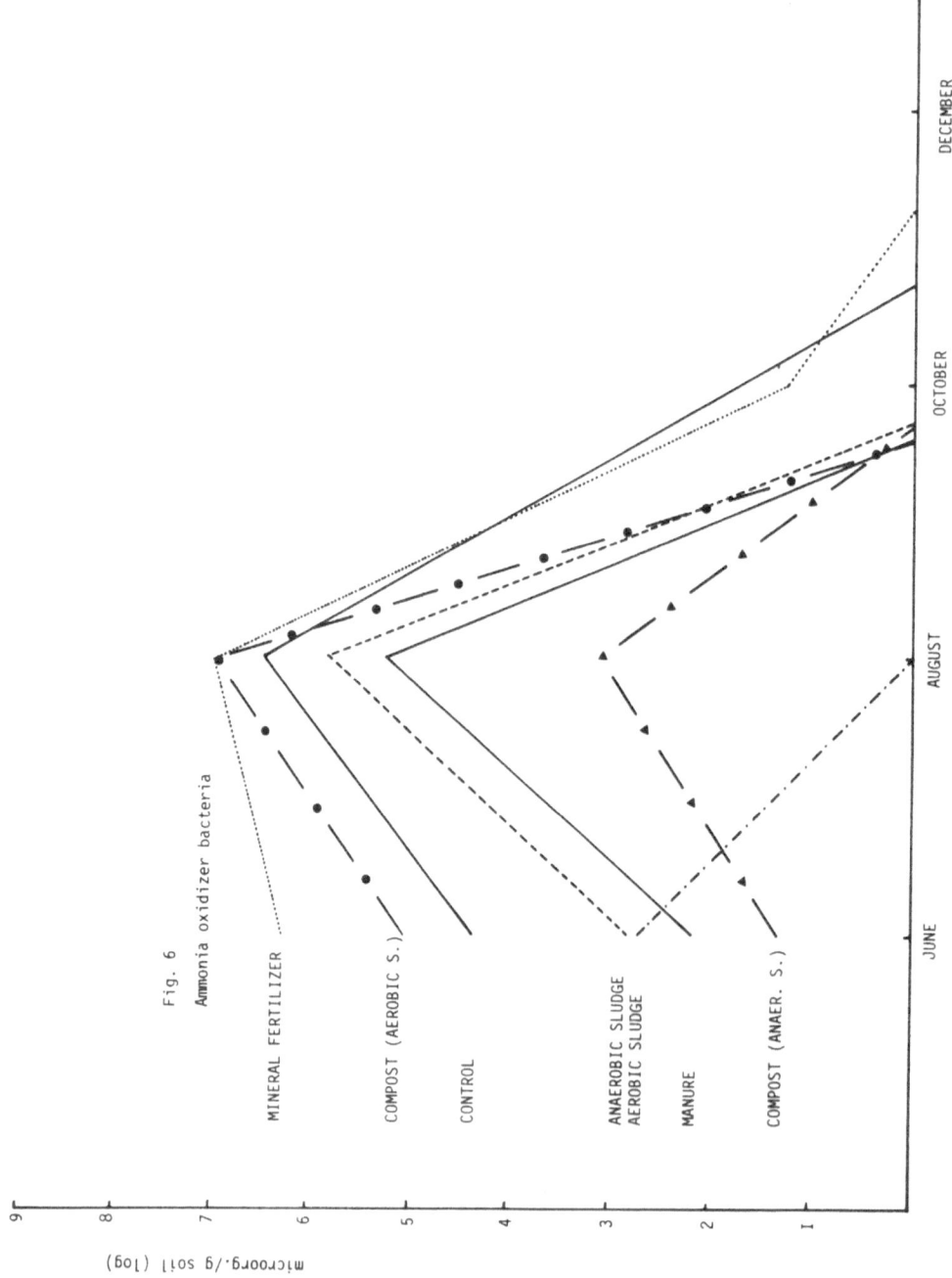

Fig. 6
Ammonia oxidizer bacteria

MINERAL FERTILIZER

COMPOST (AEROBIC S.)

CONTROL

ANAEROBIC SLUDGE
AEROBIC SLUDGE

MANURE

COMPOST (ANAER. S.)

microorg./g soil (log)

JUNE   AUGUST   OCTOBER   DECEMBER

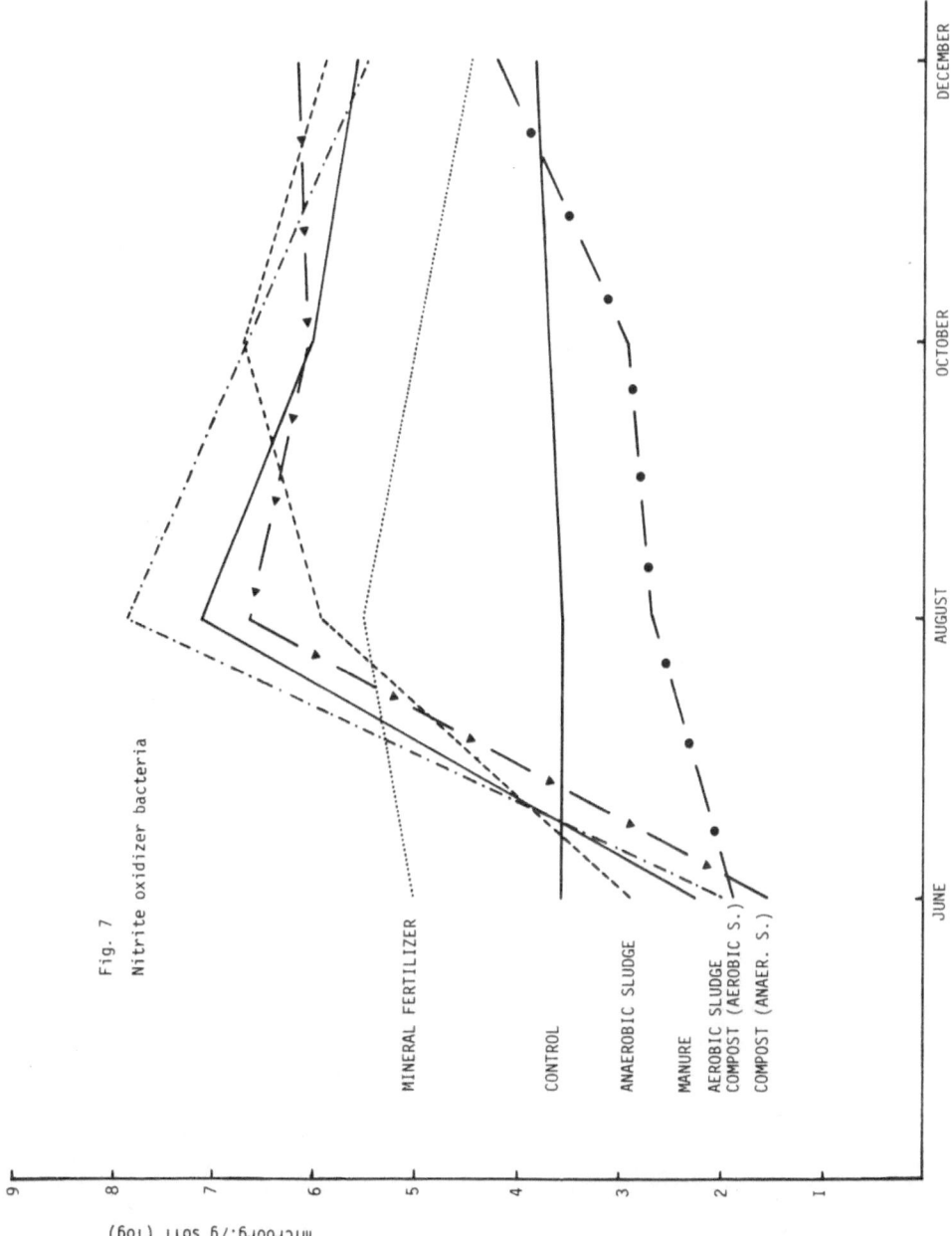

Fig. 7
Nitrite oxidizer bacteria

MINERAL FERTILIZER

CONTROL

ANAEROBIC SLUDGE

MANURE

AEROBIC SLUDGE
COMPOST (AEROBIC S.)

COMPOST (ANAER. S.)

microorg./g soil (log)

JUNE   AUGUST   OCTOBER   DECEMBER

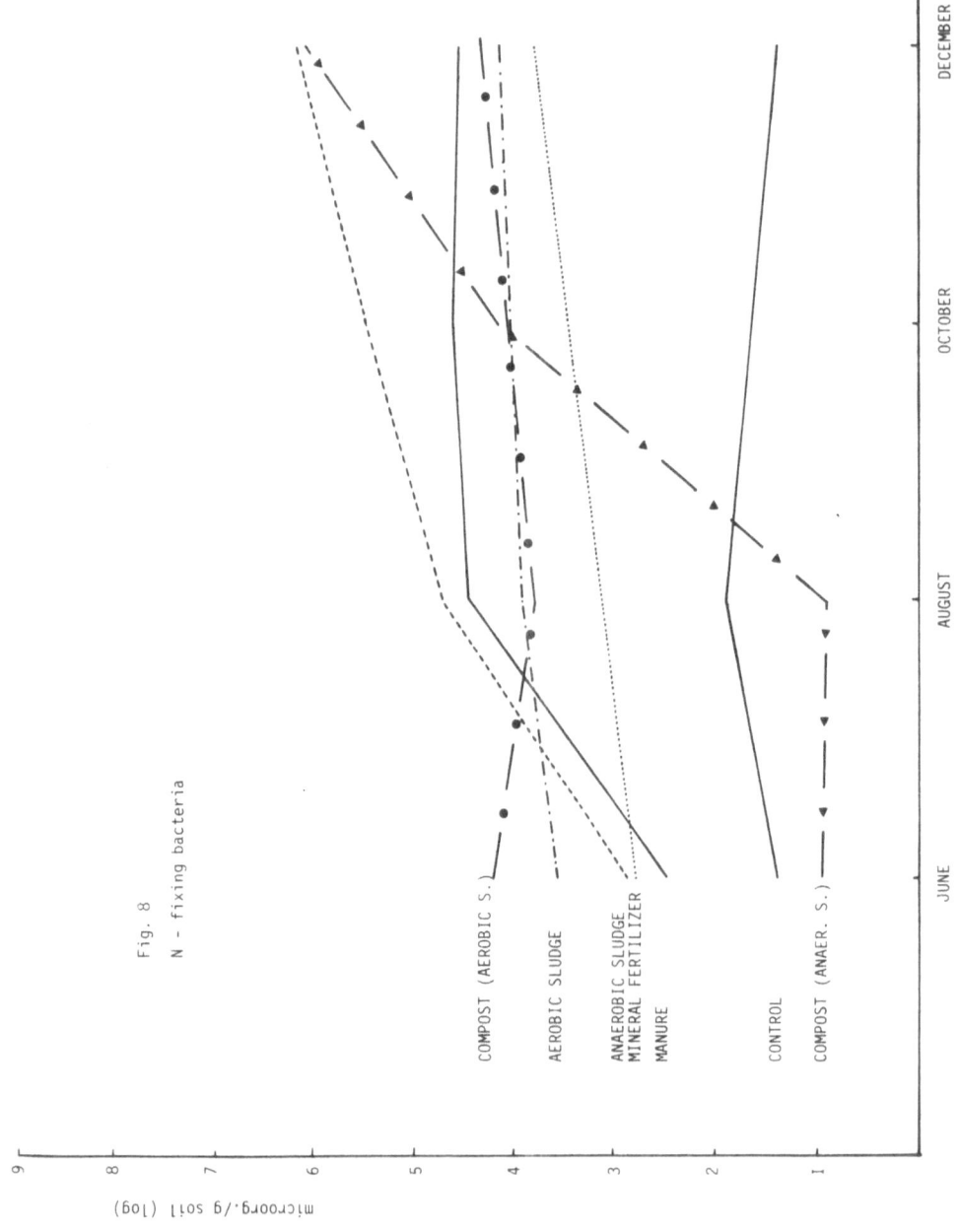

Fig. 8

N - fixing bacteria

COMPOST (AEROBIC S.)

AEROBIC SLUDGE

ANAEROBIC SLUDGE
MINERAL FERTILIZER
MANURE

CONTROL

COMPOST (ANAER. S.)

JUNE   AUGUST   OCTOBER   DECEMBER

microorg./g soil (log)

ferent treatments (Fig. 9, 10, 11). The higher total number of these fungi in anaerobic sludge treatment at the beginning of the experiment can be explained by the fact that this treatment has the highest content in organic matter.

Mycorhizal fungi. Vesicular-arbuscular mycorhizal infection in control plants reached 30 %, while in plants treated with inorganic fertilizers was only 15 %. Mycorhizal infection was absent in all the other treatments, as shown in Table 4. Numbers of Endogonaceae spores followed the same trend, decreasing from 1000/100 g dry soil in control plant rhizosphere to 150/100 g dry soil in the rhizosphere of plants treated with inorganic fertilizers.

TABLE 4 : OCCURRENCE OF MYCORHIZA IN MAIZE ROOTS AND OF ENDO-
————— GONACEAE SPORES IN ADJACENT SOIL FROM SEVEN DIFFERNT
FERTILIZING TRIALS.

Type of treatment	% infection (10 root samples)	Endogonaceae spores/ 100 g soil
Control	30 %	1,000
Mineral fertilization	15 %	150
Aerobic sludge	0	50
Anaerobic sludge	0	0
Compost (aerobic s.)	0	0
Compost (anaerobic s.)	0	50
Manure	0	0

In the other treatments Endogonaceae spores were occasionally found.

Only two genera of endophytic fungi were recovered, Glomus and Sclerocystis, with the species G. caledonius, G. mosseae, G. fasciculatus and S. rubiformis.

Our results confirm that vesicular-arbuscular mycorhiza are inhibited or reduced by high levels of available phosphate (5,11). Indeed mycorhizal infection was not observed in plants treated with high doses of phosphorus.

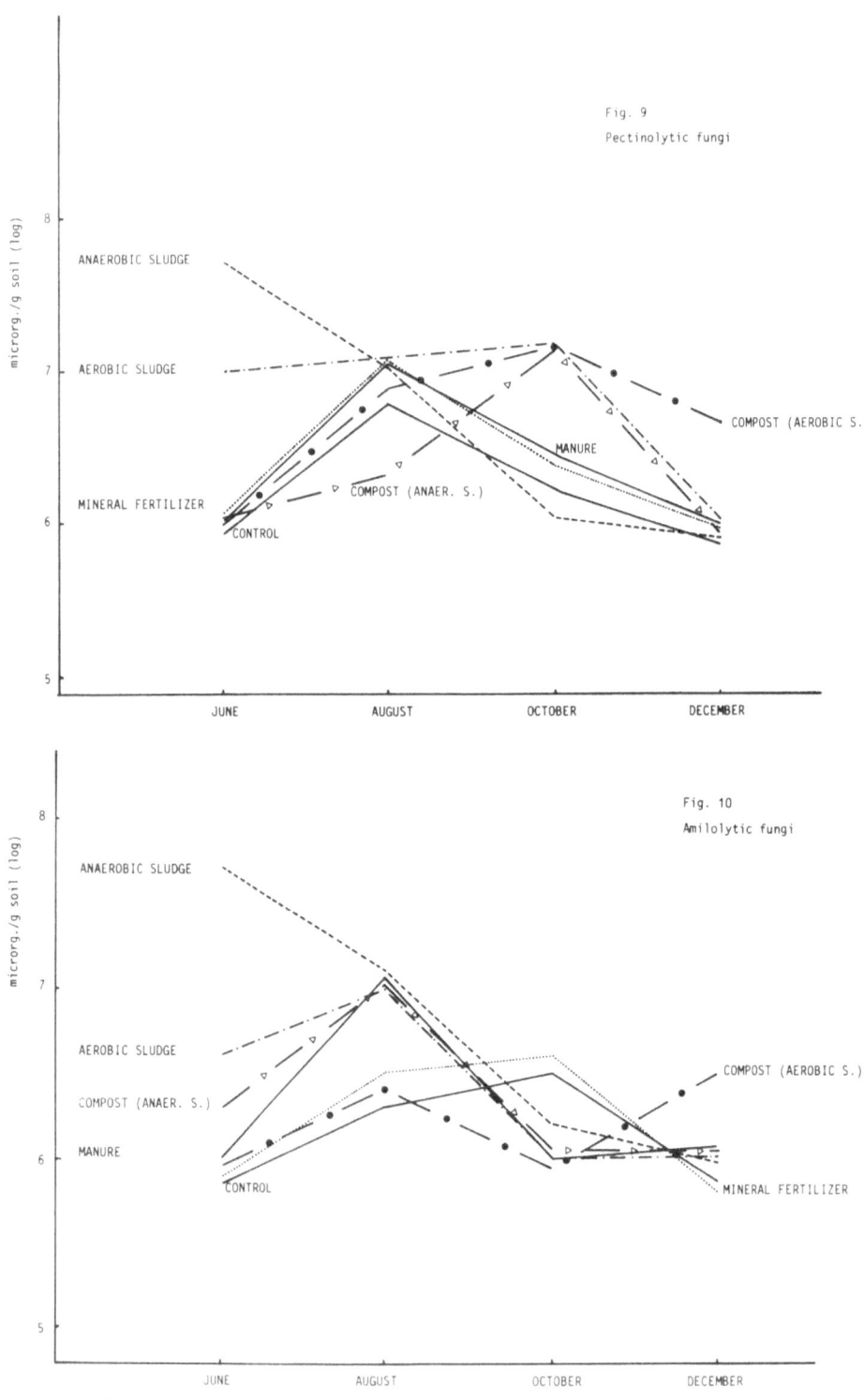

Fig. 9
Pectinolytic fungi

microrg./g soil (log)

8

ANAEROBIC SLUDGE

7

AEROBIC SLUDGE

COMPOST (AEROBIC S.)

MANURE

MINERAL FERTILIZER

COMPOST (ANAER. S.)

6

CONTROL

5

JUNE   AUGUST   OCTOBER   DECEMBER

Fig. 10
Amilolytic fungi

microrg./g soil (log)

8

ANAEROBIC SLUDGE

7

AEROBIC SLUDGE

COMPOST (ANAER. S.)

COMPOST (AEROBIC S.)

MANURE

6

CONTROL

MINERAL FERTILIZER

5

JUNE   AUGUST   OCTOBER   DECEMBER

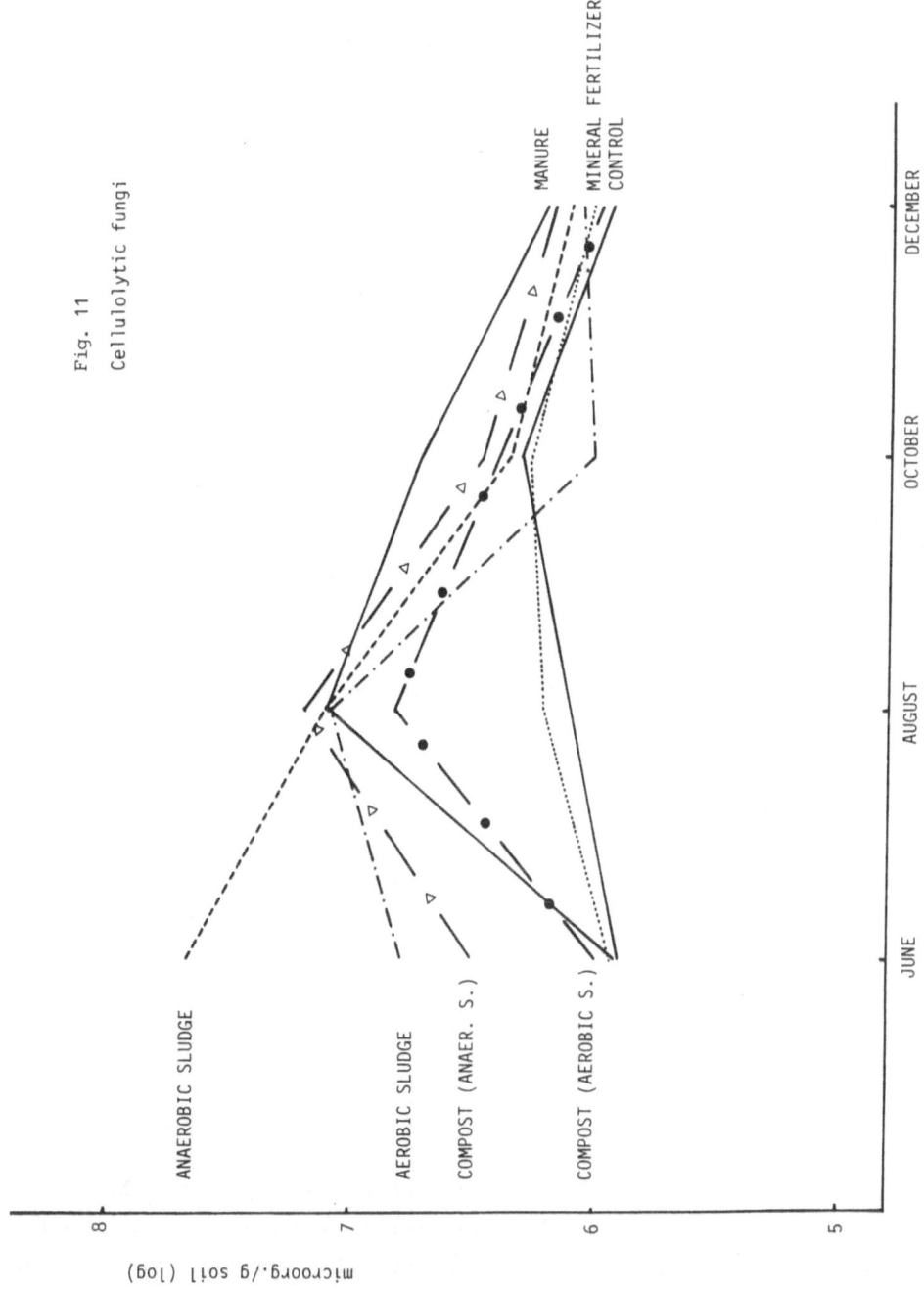

Fig. 11

Cellulolytic fungi

microorg./g soil (log)

ANAEROBIC SLUDGE

AEROBIC SLUDGE

COMPOST (ANAER. S.)

COMPOST (AEROBIC S.)

MANURE

MINERAL FERTILIZER

CONTROL

JUNE　　AUGUST　　OCTOBER　　DECEMBER

4. CONCLUSION

The results obtained in this research suggest that the rhizo-
sphere-effect is more important for microbial growth in the
soil-root interface than any other external effect, including
treatments with organic and inorganic fertilizers. The high
doses employed in our experiment may be the reason for the
slight differences observed among treatments. Differences, however, were always found in controls, which did not receive any
fertilization.

The limited variation of microbial rhizosphere population in
plants treated with different fertilizers may also be due to
soil homeostasis. As Alexander says "... when the environment
changes, self-regulatory mechanisms or homeostatic reactions
come into play to restore the relationships previously existing
among the residents and to maintain the relative constancy in
the make up of the community. Biological alterations occurr
as a result of environmental stress, but the processes contributing to homeostasis tend to restore the initial steady-state
situation" (1).

Besides these considerations, in our research we have observed
that experiments carried out in open field undergo a great number of variations which may largely affect the statistical
significance of data, expecially where microbial habitats and
ecology are concerned.

5. SUMMARY

Maize plants were fertilized with different kinds of sludge
(anaerobically digested and aerobically stabilized) and with
compost derived from sludge mixed with the organic fraction of
solid urban wastes. The effects of these treatments were studied
on soil bacteria, actinomycetes and fungi; the occurrence of

the different physiological groups is given for each treatment. Furthermore, a study was conducted on the effects of the same treatments on vesicular-arbuscular mycorhiza infections in the roots and on the total number of endophytic fungal spores in the soil.

Results of these experiments are reported and discussed in order to evaluate the implications of sludge and/or compost applications on soil microflora.

6. REFERENCES

(1) ALEXANDER M. - In "The soil-root interface" (Harley, Scott Russel Eds.). Academic Press, 1979.

(2) ALEXANDER M. - Nitrification. In "Soil Nitrogen" (W.V. Bartholomew and F.E. Clark, Eds.). Publ. N. 10, 307-343. Agr. Soc. Agron. Madison, Wisconsin, 1965.

(3) DEPARTMENT OF THE ENVIRONMENT, London. Agricultural Use of Sewage Sludge. Notes on Water Pollution, 57, June 1972.

(4) DEPARTMENT OF THE ENVIRONMENT - National Water Council, London.Report of the Working Party on the Disposal of Sewage Sludge to Land. July 1977.

(5) GERDEMANN J.W. - Vesicular-arbuscular mycorrhizae. In "The development and function of roots". (Torrey-Clarkson, Eds.), 229, 1975.

(6) GERDEMANN J.W. and T.H. NICOLSON - Spores of mycorrhizae Endogone species extracted from soil by wet sieving and decanting. Trans. Br. Mycol. Soc. 46, 235, 1963.

(7) GIOVANNETTI M. and B. MOSSE - An evaluation of technique for measuring vesicular-arbuscular mycorrhizae infection in roots. New Phytol., 84, 489, 1980.

(8) HYDE H.C. - Utilization of wastewater sludge for agricul-
     tural soil enrichment. Journal WPCF 48 (9) 2190-2197,
     1976.

(9) LOEHE, R.C. - Agricultural waste management : Problems,
     processes and approaches. Academic Press, New York and
     London, 1974.

(10) MARSHALL K.C. and M. ALEXANDER - Nitrification by
     Aspergillus flavus. J. Bacteriol., 83, 572-575, 1062.

(11) MENGE J.A., D. STEIRLE, D.J. BAGYARAI, E.L.U. JOHNSON
     and R.T. LEONARD - Phosphorus concentrations in plants
     responsible for inhibition of mycorrhiza infection.
     New. Phytol., 80 (3), 575-578, 1978.

(12) PHILLIPS J.B. and D.S. HAYMAN - Improved procedures for
     clearing roots and staining parasitic and vesicular-arbus-
     cular mycorrhizae fungi for rapid assessment of infection.
     Trans. Br. Mycol. Soc., 55, 150, 1970.

(13) POCHON J. and P. TARDIEUX - Techniques d'analyse en Mi-
     crobiologie du sol. Ed. La Tourelle, 1962.

(14) TESTER C.F., L.J. SIKORA, J.M. TAYLOR and J.F. PARR -
     Decomposition of sewage sludge in soil : Carbon and Ni-
     trogen transformation. J. Environ. Qual., 6, N.4, 459-463,
     1977.

(15) VARANKA M.W., Z.M. Zablocki and T.O. HINESLY - The ef-
     fect of digested sludge on soil biological activity.
     Journal WPCF 48 (7), 1728-1740, 1976.

(16) VERSTRAETE W. and J.P. VOETS - Soil microbial and bio-
     chemical characteristics in relation to soil management
     and fertility. Soil Biol. Biochem., 9, 253-258, 1977.

DISCUSSION

Dr WILLIAMS :   A comment. I have very little knowledge in soil microviology but is it not true that microbial populations and activity are substrate dependent ? In the case of ammonia oxidizers for example, the increase in the population count is dependent on NH_4-N concentration. When making a comparison with other fertiliser treatments - if the NH_4-N concentrations are not strictly the same as in the sludge treatments, results can be very misleading ?

Dr HALL :   You attribute the inhibition of microbial development to the very high levels of phosphorus in the sludge. Did you also apply equally high levels of inorganic phosphorus to test this ?

Dr GIONANNETTI :   We did not in this experiment, but from data existing in the litterature we can say that high levels of P inhibit vesicular-arbuscular infection.

Dr CATROUX :   Is the effect of sludge on Mycorhiza persistent ?

Dr GIONANNETTI :   In our experiment sludge applications were made in june and mycorhizal infection was measured from september to december. So, we can only say that this effect is persistent for 6 months.

Dr CATROUX :   If the effect is persistent after one sludge application, it may be of practical importance, specially in forestry ?

Dr GIONANNETTI :   We can't extend our results to forest trees
                   because they are infected by another kind of
                   mycorhizal fungi, precisely ectomycorhizae
                   caused by different fungi (basidiomycotina
                   instead of zigomycotina).

SLUDGE EFFECT ON SOIL AND RHIZOSPHERE BIOLOGICAL

ACTIVITIES

U. TOMATI, A. GRAPPELLI and E. GALLI

1. INTRODUCTION

Sewage sludge is a very valuable fertilizer and soil amendement material, since it contents N, P, organic matter and all the other constituents (as microelements, microbial metabolites, growth regulator substances, etc.). On the other hand, sludge, especially that from industrial areas, may often contain high concentrations of toxics such as heavy metals or organic polluting substances.

These substances, as well as an excess of nutrients, may create serious problems in regard to soil health.

Biological activities should be regarded as an important index for soil health. It is important to follow changes in soil microbial populations and metabolic activities resulting from slugde application in the soil as well as in the rhizosphere.

The rhizosphere could be considered as a soil differing in its physical, chemical and biological properties from that outside the root zone. The properties of the rhizosphere are extremely important for plant metabolism. Infact in this environment, the microflora role is not only limited to the mineralization of complex substances, but it is also related to the production of various active biocatalysts, including enzymes, vitamines, growth regulator substances, antibiotics, toxins and many other compounds which play a particular role in plant growth and development.

The agronomic value of sludge could be distinguished as a long term fertilizing value, dependent on storage of nutrients released from mineralization, and an actual fertilizing value, dependent on nutrients immediately available for the plants.

Soil and rhizosphere activity could be considered as indexes of these two kinds of soil fertility.

Infact, soil microflora and soil activities may supply some informations about future soil fertility, since their role in metabolizing the soil complex substances. Rhizosphere microflora and rhizosphere activities may also supply useful informations about the global metabolism in the plant-soil system. Here, the interaction microorganisms-root is a strongly conditioning factor for plant growth and development, nutrients and water uptake being regulated from many microbial products.

With the aim to investigate soil and rhizosphere activities in relation to their agricultural meaning, pot trials on maize and field trials on wheat and maize were carried out.

Soil has been treated with different amounts of sludge in order to evaluate, if any, harmful effects dependent from toxics or overfertilization.

2. POT TRIALS
   ‾‾‾‾‾‾‾‾‾

Experiments were carried out on maize (Zea mays Saturno TV 34) sown and raised in a controlled environment (temperature : 30°/20°C day/night; relative humidity : 60 %; photoperiod : 14 h; irradiation intensity : 20,000 lux). The plants were grown in plastic pots containing 1 kg of dry soil (clay : turf: silt 1:1:2).

The soil pH was 6.5. Different amounts of domestic sludge (4-8-16 g dry weight/kg of soil) were mixed with the soil and conditioned for 15 days, without any mineral.

In this series, the rhizosphere activities were measured.
In separate series, without plants, the soil activities were
analyzed.

Sludge composition (%)

H_2O	C_{tot}	C_{org}	C_{inor}	N_{tot}	P_2O_5 tot
73.39	30.21	29.64	0.57	0.93	1.40

Soil and soil rhizosphere were taken after 7, 12, 21, 27 days
from the sowing.

Experiments were replicated 3 times and the variation between
the replicates appeared to be less than 10 %.

In the soil and in the rhizosphere soil, the following measurements were carried out :

- microflora growth (bacteria, actinomycetes, fungi)
- oxygen consumption
- protease activity, nitrification and denitrification
- NO_3^- and PO_4^- content
- acid phosphatase activity
- heteroauxins production (expressed as IAA production).

3. RESULTS

The following symbols were used for the graphs :

□----□ Mineral
○----○ Soil without any fertilizer
△----△ 4 g of sludge (dry weight/kg dry soil)
▲----▲ 8 g of sludge (dry weight/kg dry soil)
●----● 16 g of sludge (dry weight/kg dry soil)

Microflora growth and oxygen consumption

Microflora growth, especially bacteria, depends largely on the amounts of supplied sludge. However, microflora growth is more greatly enhanced in the rhizosphere than in the soil.

The more rapid development of microorganisms shows that the microorganism-plant interaction may begin immediately after sowing. As the root system develops and extends in contact with the soil, soil microorganisms in the immediate vicinity are benefically influenced. Their growth is probably selective and enhanced to initiate the rhizosphere effect.

Oxygen consumption is very similar to that of the micro-flora, especially bacteria, and may be considered as an index of soil metabolism efficiency. The greater oxygen consumption in the rhizosphere may be also related to both an enhanced microbial population and greater stimulated activities.

Protease activity

Protease activity is largely dependent from the supplied amounts of sludge. In the soil, it quickly decrease, whereas in the rhizosphere a quite strong activity always remains.

The high protease activity recorded in untreated rhizosphere soil is an index that the protease activity is also dependent from root exogenous proteolytic enzymes. This fact usually is beneficial for the root environment because it permits a more continous mineralization of N-containing substances.

Nitrification and denitrification

The large oxygen consumption depends on the biodegradation of organic compounts. As a result of this, a considerable altera-tion in the ratio aerobic/anaerobic microorganisms may occur.

MICROBIAL GROWTH

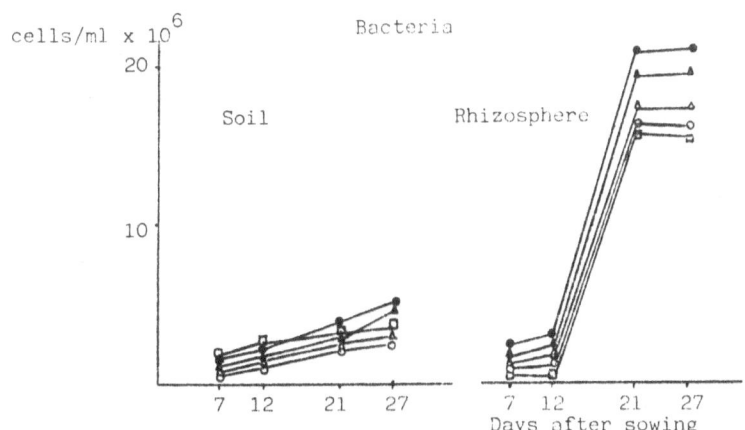

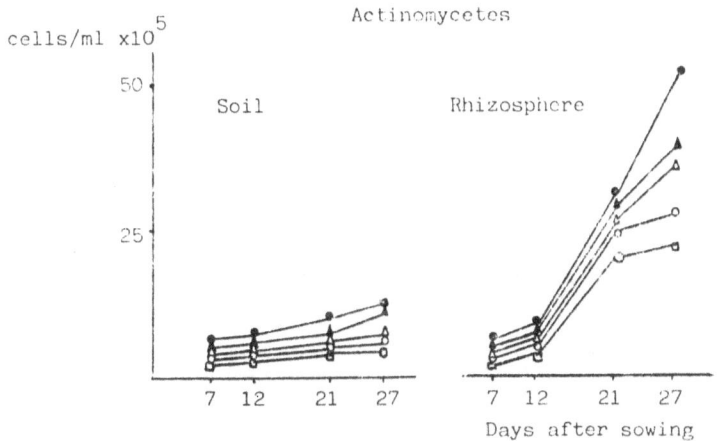

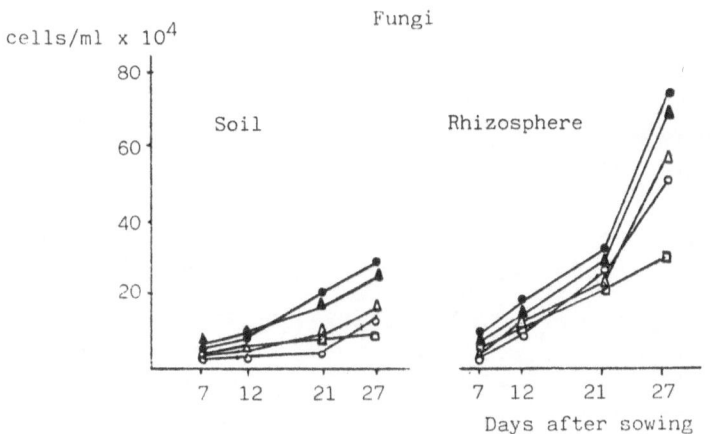

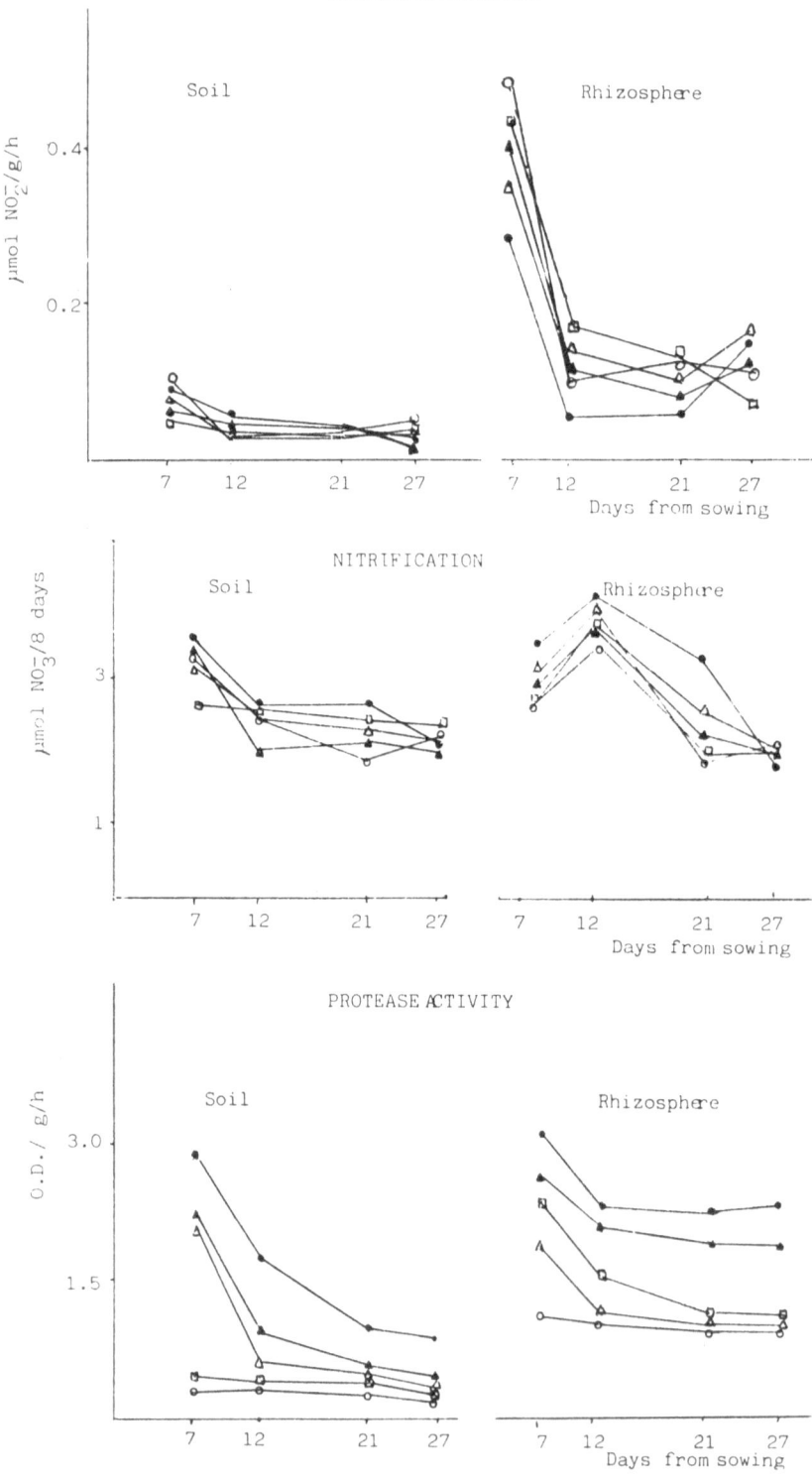

The strong oxygen consumption suggests that the ratio aerobic/
anaerobic favours aerobic microorganisms.

The behaviour of denitrification and nitrification seems to
confirm this alteration. The denitrification is more depressed
in the rhizosphere than in the soil. No evident differences are
detected for the nitrification both in the soil and in the
rhizosphere.

Nitrate content

Nitrate content both in the soil and in the rhizosphere supply
further information about nitrogen mineralization and nitrate
utilization.

At the end of the experiment, nitrate concentration in the soil
is higher than in the corresponding rhizosphere. The enhanced
nitrate content in the soil is an index of good sludge mineralization. The lower nitrate concentration, recorded in the
rhizosphere, indicates its utilization both by microorganisms
and plants.

In agronomic terms, soil activities and nitrate content in soil
could be considered as an index of the future soil fertility,
whereas higher rhizosphere activities and nitrate consumption
could be an index of actual fertility.

Acid phosphatase activity

Acid phosphatase activity is very similar to that one of
bacterial growth and seems to confirm the hypothesis of a cor-
relation between bacterial growth and acid phosphatase activity.

As all the other activities, acid phosphatase activity is
largely dependent from the sludge amounts and is more enhanced
in the rhizosphere. The strong rhizosphere activity may be

NITRATE CONCENTRATION (μmol NO_3^-/g dry weight)

SOIL	Days after sowing				
	0	7	12	21	27
Mineral	12.3	19.0	19.4	27.6	23.6
Soil	7.1	15.6	17.3	24.5	16.0
Soil + 4 g of sludge (d.w.)	10.1	19.2	13.0	23.1	18.5
Soil + 8 g of sludge (d.w.)	11.0	16.2	19.0	18.7	20.6
Soil + 16 g of sludge (d.w.)	10.1	14.4	15.3	18.2	27.8
RHIZOSPHERE					
Mineral	12.3	20.6	20.4	21.4	17.3
Soil	7.1	18.0	15.8	11.0	13.8
Soil + 4 g of sludge (d.w.)	10.1	18.5	21.0	15.8	15.0
Soil + 8 g of sludge (d.w.)	11.0	19.2	18.0	21.7	19.7
Soil + 16 g of sludge (d.w.)	10.1	15.8	16.2	14.9	12.6

Error : $\pm$ 0.3

PHOSPHATE CONCENTRATION (μmol P/g dry weight)

SOIL	Days after sowing				
	0	7	12	21	27
Mineral	2.6	2.8	2.8	3.1	4.1
Soil	3.0	3.0	3.5	2.9	3.0
Soil + 4 g of sludge (d.w.)	3.0	2.8	3.0	4.3	4.4
Soil + 8 g of sludge (d.w.)	2.2	4.2	4.3	4.4	5.1
Soil + 16 g of sludge (d.w.)	2.9	3.6	3.8	4.4	5.2
RHIZOSPHERE					
Mineral	2.6	3.2	2.8	4.5	3.9
Soil	3.0	3.4	3.2	3.3	3.4
Soil + 4 g of sludge (d.w.)	3.0	3.5	4.0	4.3	2.7
Soil + 8 g of sludge (d.w.)	2.2	3.9	3.9	10.5	5.6
Soil + 16 g of sludge (d.w.)	2.9	7.0	7.0	9.1	4.9

Error : $\pm$ 0.3.

explained by the hight phosphate amount required by microflora
and plant development.

Phosphate content

Phosphate content increased both in soil and in the rhizosphere
even if data shows a large phosphate consumption in the rhizosphere.

Hormonal production

Sludge is not only a source of macro and micronutrients, but
it must also be considered in view of its content in biocatalysts, as aminoacids, vitamines, enzymes and growth regulator
substances.

Growth regulator substances are very important in plant metabolism because thery are able to influence nutrient uptake and
plant metabolism, especially during the first stage of plant
development.

Therefore, we tested hormon production (expressed as IAA) in
relation to different sludge amounts.

Results show that heteroauxin production is low in the soil and
very high in the rhizosphere. This fact is of particular importance because metabolic activities in the plants, such as
water and ion absorption, taking place in the rhizosphere, are
influenced by growth regulators.

OXIGEN CONSUMPTION

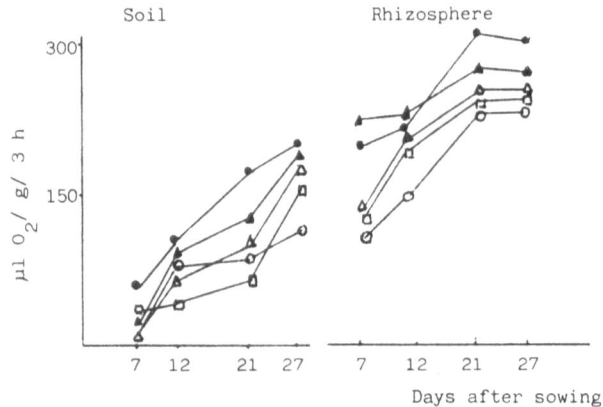

ACID PHOSPHATASE ACTIVITY

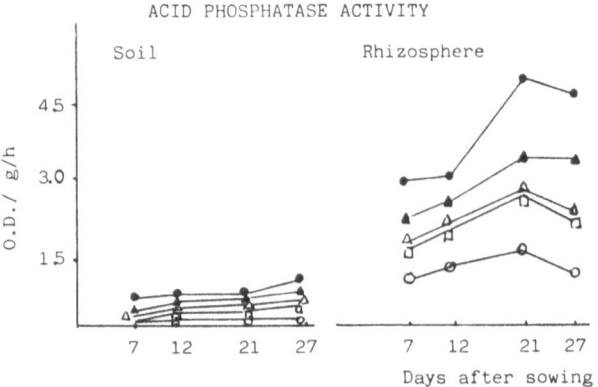

AUXINS PRODUCTION

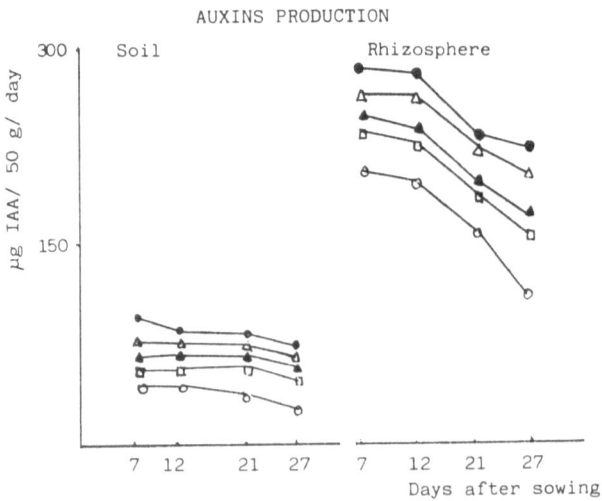

4. FIELD EXPERIMENTS

Experiments have to be carried out for 8 years for the maize
and 4 years for the wheat in rotation to sugar beet. Experiments started in 1978 and will finish in 1985.

Aerobic or anaerobic sludge from municipal plants are used.
Sludge is distributed by spreading on 500 square meters plots.
The soil is a sandy-loam soil; pH is 6.5. Two doses of both
sludge are early supplied (10 tons dry matter/hectar and 30
tons dry matter/hectar).

Biological activities are tested during the significant moments
of vegetative cycle of the plants.
For the maize, samples are taken out at germination, emergence,
blooming, ripening; for the wheat at germination, tillering,
emergence and blooming.

Sludge composition (%)

	d.m.	ashes	C_{org}	N_{tot}	P_2O_5
Aerobic sludge	2.1	40.0	27.54	3.93	7.83
Anaerobic sludge	9.9	29.7	37.99	2.99	2.53

At present, our project result is only based on 3 years for the
maize and 2 years for the wheat. So, as we cannot came to a
definite conclusion, we can give you an idea about the relationship between soil and soil rhizosphere activities and, if
possible, on their agricultural meanings.

- As the pot trials, also field experiments show that all
  activity and microflora development are higher in the
  rhizosphere;
- activities are dependent from sludge amounts;
- the highest dose shows harmful effects related to anfertili-
  zation;

In this case, several activities in the soil and rhizosphere are depressed.

- there is some evidence for a relationship among rhizosphere activities, plant development and crop yield.

5. REFERENCES

(1) CHAUSSOD R. and GERMON Y.C. : Mineralisation dans le sol de l'azote de différentes boues de stations d'epuration. C.R. des Sciences de l'Academie d'Agriculture de France, vol. LXIII, 525-531; (1977).

(2) COSGROVE D.J. and al. : Inositol phosphate phosphatases of microbiological origin. The isolation of soil bacteria having inositol phosphate phosphatase activity. Aust. J. Biol. Sci. 23, 339-343; (1970).

(3) FAUVEL B. and ROUQUEROL T. : The phosphatase test considered as an index of soil activity and evolution. Revue Ecol. Biol. Sol. 7, 393-406; (1970).

(4) FRIED M. and BROESHART H. : "The soil-Plant System"; Academic Press, (1967).

(5) GERMON J.C. : Effets d'épandages répétés d'eaux résiduaires de conserveries sur la microflore du sol. C.R. des Sciences de l'Académie d'Agriculture de France, vol. LXIII, 516-524; (1977).

(6) VARANKA M.W., ZABLOCKI Z.M. and HINESLY T.D. : The effect of digested sludge on soil biological activity. Journal WPCF vol. 48 n° 7, 1728-1729, July 1976.

(7) NANNIPIERI P., JOHNSON R.L. and PAUL E.A. : Criteria for measurement of microbial growth and activity in soil. Soil Biol. Biochem. vol. 10, 223-229; (1978).

(8) NANNIPIERI P., PEDRAZZINI F., ARCARA P.G. and PIOVANALE C. :
Changes in aminoacids, enzyme activities, and biomasses
during soil microbial growth.
Soil Science vol. 127 n° 1, 26-34; (1979).

(9) ROVIRA A.D. and McDOUGALL B.M. : Microbial and biochemical
aspects of the rhizosphere.
Soil Biochem. 1, 417-463; (1967).

(10) TATE R.L. III : Microbial activity in organic soils affected
by soil depth and crop.
Applied and Environmental Microbiology, 1085-1090, June
1979.

(11) TECHNICAL BULLETIN : Municipal sludge management, Environmental Factors. EPS 430/9-77-004, October 1977.

(12) VERSTRAETE W. and VOETS J.P. : Soil microbial and biochemical characteristics in relation to soil management
and fertility. Soil Biol. biochem. vol. 9, 235-238 (1977).

(13) VRANY J. : Oxygen consumption as a criterion of actual
soil metabolic activity in the root proximity. Studies
about humus. BRNO CZECHOSLOVAKIA Humus et Planta VII (1979).

DISCUSSION

Dr CATROUX :   Regarding your results, Dr TOMATI, it is
difficult to distinguish between a direct effect via the plant with an improved growth
with sludge treatment ?

Dr TOMATI :   It is not easy because of the relationships in
the plantsoil system are very complex. However,
the experiments without any sludge can give
us an idea of the plant influence on rhizosphere activities.
I think, however, that it is important to consider the activities in the wole plant-soil
system as influenced by sludge.

SOME BIOLOGICAL PROPERTIES IN SEWAGE SLUDGE AND SEWAGE

TREATED SOILS

E. VIGERUST

1. EFFECT OF RAW SLUDGE

Only a small part of the norwegian sludge is biologically
stabilized. When we apply undigested sludge to the soil a very
quick decomposition will take place in the soil. We can talk
about "soil stabilization". The microbial activity will also
depend on the amount of sludge applied. Lime stabilized
sludge is not biologically stabilized either and will have
nearly the same effects in the soil as raw sludge.

The high microbiological activity which may cause reduced
growth, is more pronounced in pot trials than in the field.
A typical result from pot trial is expressed in fig. 1. The raw
sludge is stored for about one year and therefore partly decomposed. Grass is here sawn in pure sludge. During the summer
season there has been 4 cuttings.

The composted sludge has a relatively high content of easy
available nitrogen. The degradation of the organic material is
slow and the liberation of nitrogen is little. Partly decom-
posted sludge has higher nitrogen content. The activitiy will
however cause growth depression. This may be due to the toxic
effects of some intermediate products.

After surviving the first 1-1 1/2 months, the grass grows
excellently due to the high content of the liberated, readily
available nitrogen.

MARTINSEN, in his pot trials (1976), mixed sludge with soil at different times before sowing barley in the pots, which were stored at different times, temporarily either stored outdoors or indoors until the sowing, in may, see table 1.

TABLE 1 : EFFECT OF RAW SLUDGE 80 TONS D.M./HECTARE   APPLIED
————————   TO SOIL AT DIFFERENT TIME BEFORE SOWING.6.5. . YIELD
BARLEY-GRAIN G PR POT (MARTINSEN, 1976).

	g N pr pot		
	0	0,3	0,6
Control	3.3	5.9	13.1
Raw sludge mixed 6.5.	1.0	5.7	8.9
Sludge compost mixed 6.5.	11.3	13.3	13.8
Sludge appl. 1.4. Stored indoor	2.7	5.3	5.9
Sludge appl. 31.1 Stored indoor	8.4	9.8	9.5
Sludge appl. 2.11. Stored indoor	9.6	12.4	13.5
Stored outdoor	0.4	4.4	6.2

Fig. 1 : Yields of grass at 4 cuttings in pot trials.

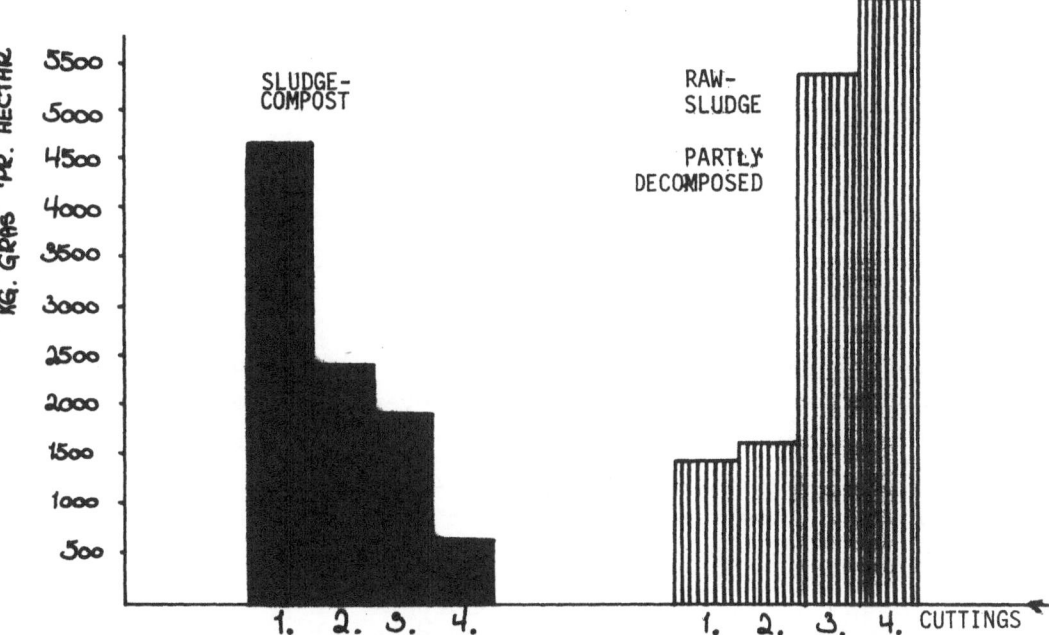

Well decomposed sludge will give high yield.

Application of fresh sludge just before sowing has reduced the
yield. The longer the pots are stored indoors before sawing,
the better is the sludge decomposed and the negative effect on
yield is less.

Application of nitrogen has to same extent compensated for the
negative effect of undecomposed sludge. Even though the sludge
material is rich in N, the microorganisms may fix such large
amounts of the nitrogen during rapid degradation processes that
the plants may not receive sufficient amounts of the element.

This however does not explain the whole depression effect of
raw sludge.

In the field the depressive effect of undigested sludge is much less than in pot trials. MARTINSEN (1976) found following effects of raw sludge applied in the spring, kg gram.

Sludge application tons D.M./hectare

	0	30	60
1 year (4 trials)	223	262	293
2 years after effect (5 trials)	276	309	332
3 years after effect (2 trials)	240	292	320

There are significant positive effects of **raw sludge** in three years. The response first year is however **less** than usual for digested sludge. After heavy applications of **raw** sludge the plants will show a slight growth depression a period after germination. This damage seems to be **most pronouced** in dry years.

The difference in the effects obtained in pot trials and field trials is probably due to the rapid decomposition in the pots where the temperature is higher.

2. CHANGES IN pH

The effects of sludge application on soil pH are of special interest when considering the availability of heavy metals. CHANEY et al (1978) has found that sludge application may cause a certain drop in soil pH and this may cause phytotoxicity by heavy metals.

The changes in pH can be related to microbial activity especially to the conversion of organic N to inorganic N. In the first stage organic N in the sludge is converted to NH_4-N (ammonification). This leads to a slight increase in pH. When the conversion of NH_4-N to NO_3-N (nitrification) takes place, the pH drops.

In pot trials we, among other things, studied the nitrogen-turnover in combination with and without drainage of the pots and with and without vegetation. Nitrogen-compounds pH etc are determined at different times. The results are shown in table 2.

TABLE 2 : pH IN DIFFERENT DEPTHS AND MEASURED AT DIFFERENT TIMES IN SLUDGE (100 1 pots)

			1979 May	1979 Oct.	1980 Oct.	1981 May
Drainage	Vegetation	0-10 cm	7.2	5.6	5.6	5.8
		20-30 "	7.2	7.6	5.4	5.1
		40-50 "	7.2	7.9	5.4	5.4
Drainage	No vegetation	0-10 cm	7.2	7.0	5.1	5.4
		20-30 "	7.2	7.7	5.3	5.7
		40-50 "	7.2	8.0	6.2	6.2
No drainage	Vegetation	0-10 cm	7.2	6.8	6.9	5.5
		20-30 "	7.2	7.5	6.8	5.0
		40-50 "	7.2	7.8	6.9	5.0
No drainage	No vegetation	0-10 cm	7.2	7.0	7.6	7.2
		20-30 "	7.2	7.5	7.6	7.2
		40-50 "	7.2	7.5	7.7	7.1

There was here a clear accordance between the pH-value and the amount of NO_3-N in the sludge. pH will drop in those levels where nitrate is produced and rise again when nitrate is taken up by the plants or leached out.

In a field trial where bushes were planted in a 40 cm layer of pure sludge we found this development in pH-values :

	1978 Oct.(start)	1979 May	1979 Oct.	1980 May	1980 Oct.	1981 May
0 - 15 cm	6.8	6.8	5.3	5.5	5.9	6.2
15 - 30 cm	6.8	6.5	4.7	5.6	5.6	6.0

During the first summer the pH in the sludge dropped drastically, however, rising again as time went on.

3. SURVIVAL OF POTATO-EELWORM

In Norway sludge is not considered to be a carrier of infectious plant diseases. An exception, however, are certain species of potato nematodes (Globodera rostoshiensin and Globodera pallida). These are rootparasites on potatoes. In Norway only some agricultural land is infected with Potato-Eelworm. Large areas are still free from it. Therefore we have certain restrictions in order to avoid the spreading of this disease.

Sewage sludge is one possibility in spreading viable cysts to new areas. Cysts of Potato-Eelworm can follow the soil on the potato-tubers and after wahing they are transported by the wastewater. In this way they may end up in the sewage sludge.

In Norway MUNKEBY (1978) has carried out investigations about the survival of cysts in sludge after different sludge-treatment.
Following conclusions can be drawn from these investigations :

1) Anaerobic digestion, realized in the right and safe way, will kill the cysts nearly up to 100 percent.

2) Composting which gives temperatureup to 40°C or more will give sludge free from viable cysts.

3) Aerobic digestion will reduce the number of cysts, but this treatment alone will not be a safety measure.

4) Lime-sludge with pH 10 or more in 14 days will kill the cysts.

5) During storage the number of viable cysts will decrease. In the outer surface layer however they might survive up till 4 months. Beneath the surface all cysts will be killed after 1-2 weeks. When considering the effect of storage the method of treatment at the sewage plant is of importance.

4. REFERENCES

(1) CHANEY, R.L., HUNDEMANN, W.T., PALMER, R.J., SMALL, R.J., WHITE, M.C. and DECKER, A.M. 1978. Plant accumulation of heavy metals and phytotoxicity resulting from utilization of sewage sludge and sludge composts on cropland. Proc. 1977 Nat. Conf. Composting of Municipal Residues and Sludges pp. 86-97. Information Trans. Inc. Rockwill, Md. 1978.

(2) MARTINSEN, J., 1976. Bruk av septiktankslam og råslam ved dyrking av korn. PRA-rapport.

(3) MUNKEBY, O. 1979. Potetcystenematoder i fast og flytende avfall. Stensiltrykk 10 s.

LIST OF PARTICIPANTS

BARIDEAU, L.

Faculté des Sciences Agronomiques de l'Etat
B - 5800 GEMBLOUX

BECK, J.

Bayerische Landesanstalt für Bodenkultur und
Pflanzenbau
Menzingerstr. 54
D - 8000 MUENCHEN 19

BERGLUND, S.

The National Swedish Environmental Protection
Board
Box 1302
Smidesvägen 5
S - 17125 SOLNA

BORCHERT, H.

Bayerische Landsanstalt fü Bodenkultur und
Pflanzenbau
Menzingerstr. 54
D - 8000 MUENCHEN 19

BORTLISZ, J.

Lippeverband
Kronprinzenstr. 24
D - 43 ESSEN

CATROUX, G.

Laboratoire de Microbiologie des Sols
INRA de Dijon, B.V. 1540
7, rue Sully
F - 21034 DIJON CEDEX

CHAUSSOD, R.,

Laboratoire de Microbiologie des Sols
INRA de Dijon, B.V. 1540
7, rue Sully
F - 21034 DIJON CEDEX

COPPPOLA, S.,

Istituto di Microbiologia Agraria
Università di Napoli
I - 80055 PORTICI-NAPOLI

DANNEBERG, O.H.

Oesterreichische Studiengesellschaft für Atom-
energie GmbH
Lenaugasse 10
A - 1082 WIEN

DE HAAN, S.

Instituut voor Bodemvruchtbaarheid
Oosterweg 92
NL - HAREN

DEHANDTSCHUTTER, J.

Directorate-General for Agriculture
Commission of the European Communities
200, rue de la Loi
B - 1049 BRUSSELS

DIEZ, Th.	Bayerische Landesanstalt für Bodenkultur und Pflanzenbau Menzingerstr. 54 D - 8000 MUENCHEN 19
FURRER, O.J.	Forschungsanstalt für Agrikulturchemie und Umwelthygiene Schwarzenburgerstrasse 155 CH - 3097 LIEBEFELD
GIOVANNETTI, M.	Centro di Studi per la Microbiologia del Suolo CNR I - 56100 PISA
GUIDI, G.	Laboratoria per la Chimica del Terreno CNR Via Corridoni 78 I - 56100 PISA
HALL, J.E.	Water Research Centre Stevenage Laboratory Elder Way UK - STEVENAGE, Herts SG1 1TH
KOSKELA, I.	Agricultural Research Center Department of Agricultural Chemistry and Physics P.O. Box 18 SF - 01301 VANTAA 30
L'HERMITE, P.	Directorate-General for Science, Research and Development Commission of the European Communities 200, rue de la Loi B - 1049 BRUSSELS
MOREL, J.L.	Ecole Nationale Supérieure d'Agronomie et des Industries Alimentaires 38, rue Sainte Catherine F - 54000 NANCY
SUESS, E.	Bayerische Landesanstalt für Bodenkultur und Pflanzenbau Menzingerstr. 54 D - 8000 MUENCHEN 19
TOMATI, U.	Istituto di Radiobiochimica ed Ecofisiologia Vegetale Consiglio Nazionale delle Ricerche Area delle Ricerche di Roma Via Salaria - Km 20300 I - 00016 MONTEROTONDO SCALO (ROMA)
VIGERUST, E.	Institute for Jordkultur N - 1432 AS NLH
WILLIAMS, J.H.	Ministry of Agriculture Fisheries and Food Woodthorne UK - WOLVERHAMPTON, Staffs. WV6 8TQ

INDEX OF AUTHORS